Making Ecologies on Private Land

"This book provides excellent insights into how landscapes are never still and what this means for people and environment in contemporary rural Australia. It shows how ecological processes, property arrangements, and the experiences of new landowners challenge ideas about nature and the conservation practices that flow from them. With a strong foundation in cultural environmental research, it incisively demonstrates how critical environmental social science can reframe and inform conservation policy and practice."

—Associate Professor Nicholas Gill, *University of Wollongong*

"Nature conservation on private land is critical for the long-term survival for so many of our species and ecosystems. Yet our understanding of what motivates private landholders to undertake certain conservation activities is still in its infancy. This book provides an important advance in our understanding of the different and often complex social drivers for this critical activity."

—Dr. James Fitzsimons, *The Nature Conservancy, Australia*

Benjamin Cooke · Ruth Lane

Making Ecologies on Private Land

Conservation Practice in Rural-Amenity Landscapes

palgrave
macmillan

Benjamin Cooke
School of Global, Urban and Social Studies
RMIT University
Melbourne, VIC, Australia

Ruth Lane
School of Social Sciences
Monash University
Melbourne, VIC, Australia

ISBN 978-3-030-31217-6 ISBN 978-3-030-31218-3 (eBook)
https://doi.org/10.1007/978-3-030-31218-3

This Palgrave Pivot imprint is published by the registered company Springer Nature Switzerland AG
The registered company address is: Gewerbestrasse 11, 6330 Cham, Switzerland

Acknowledgements

The authors would like to acknowledge the people of the Woi wurrung and Boon wurrung language groups of the eastern Kulin Nations on whose unceded lands this work was conducted, respectfully recognising Elders and Ancestors past and present. We would like to thank the landholders who gave their time to chat about conservation and show Ben around their properties, sharing with us the trials and tribulations of caring for ecologies. These landholders and their experiences are the heart of the book. We are extremely grateful to Ingereth Macfarlane for her crucial editorial insights on a draft version, which improved the manuscript greatly. Thanks also to Gabriella Corbo-Perkins for her excellent research assistance on Chapter 6 and to Nicholas Cooke for his help with figure design.

Ben was supported in the writing and research for this book by research leave from the School of Global, Urban and Social Studies, as well as an ARC Discovery Project Grant DP180103118. The 2010 round of empirical research was supported by an ARC Linkage Grant LP0882780, and the 2016 empirical research was partly supported by a grant from the Department of Environment, Land, Water and Planning (DELWP). Finally, Ben would like to thank Rosie and Sonny for their love, and co-author Ruth for her ongoing mentorship and support.

Sections of This Book Draw from the Following Published Material

Chapters 1 and 7 draw on themes from the journal article:
Cooke, Benjamin, and Ruth Lane. 2018. "Plant–Human Commoning: Navigating Enclosure, Neoliberal Conservation, and Plant Mobility in Exurban Landscapes." Annals of the American Association of Geographers 108 (6): 1715–1731.

Chapter 3 is based on the journal article:
Cooke, Benjamin, and Ruth Lane. 2015. "How Do Amenity Migrants Learn to Be Environmental Stewards of Rural Landscapes?" *Landscape and Urban Planning* 134: 43–52.

Chapter 4 is based on the journal article:
Cooke, Benjamin, and Ruth Lane. 2015. "Re-Thinking Rural-Amenity Ecologies for Environmental Management in the Anthropocene." *Geoforum* 65.

Chapter 6 is based on the journal article:
Cooke, Benjamin, and Gabriella Corbo-Perkins. 2018. "Co-Opting and Resisting Market Based Instruments for Private Land Conservation." *Land Use Policy* 70. Elsevier: 172–181.

Contents

About the Authors

Dr. Benjamin Cooke is a human geographer and Senior Lecturer in the School of Global, Urban and Social Studies, and the Centre for Urban Research, at RMIT University. Benjamin teaches courses on environmental governance and planning and applied research. Benjamin's research centres on the social and critical dimensions of nature conservation, which encompasses private land conservation, rural-amenity landscapes, green cities and urban greening, property rights and socio-ecological commons/commoning. He is currently undertaking an ARC research project on the rise and enabling of private protected areas in Australia (DP180103118).

Dr. Ruth Lane is a human geographer and a Senior Lecturer in the School of Social Sciences at Monash University where she teaches courses on environmental governance and impact assessment. Her research focuses on the intersections between social change, environmental degradation and environmental governance, at scales ranging from the household to the region and the nation. She has examined this through research on the social and cultural aspects of land-use change in rural and regional Australia, and on the consumption of goods and materials, and associated issues of waste, reuse and materials recycling in urban environments.

List of Figures

Chapter 1

Chapter 3

Chapter 4

Chapter 5

Chapter 6

CHAPTER 1

Introduction

Abstract This chapter begins by outlining the rise of interest that private land is receiving as a land tenure for pursuing conservation. We outline the need to understand how private land conservation practices are being undertaken in the dynamic context of rural-amenity landscapes. We set up the idea of 'conservation practice' as something undertaken by humans and nonhumans in concert, rather than a solely human endeavour. We focus our attention on the agency of plants as part of conservation practice. This chapter introduces rural-amenity landscapes as a vital context in which to explore conservation practice, detailing the hinterland regions of Melbourne, Australia as a case study. We conclude by introducing the chapter topics: private property relations, experiential learning, landscape legacy, conservation covenants and market-based instruments.

Keywords Conservation practice · Private land conservation · Exurban · Rural-amenity · More-than-human

B. Cooke and R. Lane, *Making Ecologies on Private Land*,
https://doi.org/10.1007/978-3-030-31218-3_1

> I don't know what the bush should be like. I have no bloody idea what it should be like... (Liz, conservation landholder)

1 Private Land Conservation and Rural-Amenity Landscapes

This book is about the everyday practice of nature conservation on private land. While the pursuit of nature conservation has been most frequently associated with public lands, our focus is on the less familiar, more intimate and sometimes messy practices on private land. Why is this of interest or relevance to wider conservation efforts? Over the past 30 years, the governments of western liberal nations like the United States, UK and Australia (the setting for this book) have taken a much less active role in the purchase and management of terrestrial lands for conservation. While there are a range of factors at play in this shift, it is fair to say that conservation has not been spared amidst the broader retreat of government from direct investment in public goods or the management of the public estate (Kay, 2016). Indeed, justifications for reducing government support for conservation on public land have often been grounded in the argument that private individuals on private land must be the ones to lead the way.

Private land conservation has also become a focal point for conservation due to the recognition that public protected areas are not comprehensive or adequate when it comes to protecting biodiversity. With public protected areas often having been declared historically due on their limited resource utility rather than their inherent ecological characteristics, private land can be vital for protecting some of the only examples of vegetation communities or fauna habitat. Indeed, in Australia, where over 65% of the land mass is private or leasehold land, many threatened species only persist across privately owned land (Hardy, Fitzsimons, Bekessy, & Gordon, 2017).

The growing interest in private lands as a place for conservation has seen a recognition of the need to explore the perspectives and experiences of the landholders who are directly engaged in these efforts. Much of this research has focused on the motivations and values of people engaged in conservation, to understand how policy interventions could be crafted to assist these endeavours, as well as encouraging others to get involved. This is an important work. Yet, as we will argue, to understand conservation it is vital to situate and contextualise people in the ecologies they are seeking to affect. It also means acknowledging that ecologies themselves affect the people who inhabit them, given that in situ experiences inform people's understanding of how ecologies function and what is required to care for them. A sense that conservation is a 'practice'—a process of human-environment interaction over time—is necessary for deepening our understanding of private land conservation. As such, this book foregrounds these relationships between humans and nonhumans in conceptions of how conservation plays out.

The reflection from Liz that opens this chapter offers an insight into why human-environment interactions are a vital focal point. After three decades caring for the ecologies on her property in the hinterlands of Melbourne, Australia, Liz carries substantial local conservation knowledge. Her practices of planting and weeding have helped shape the landscape she inhabits, but so has the seemingly simple act of 'walking through' (Liz) those ecologies. As the inheritor of a property cleared for grazing after colonial invasion well over a century ago, Liz has interacted with and observed plants, animals, insects and soil in their efforts to cobble together a more complex ecological assemblage. In those immersive moments, Liz has seen small birds taking up residence in plants considered weeds. She has experienced the leisurely growth of trees in sandy soils, which revealed the slow timescales of regeneration in relation to a human life. These experiences have made Liz a humble practitioner—one who prefers to be hands-off where possible, due to a concern that her interventions could further slow ecological recovery. When Liz says she does not know what the bush should be, she is not suggesting she knows nothing about it. Rather, this is an acknowledgement that she is but one player amongst a host of past and present humans and nonhumans acting in this place. Her role is one of shaping the bush in concert with others, not as sole arbiter over what it will or should look like.

As we will explore throughout the chapters in this book, a focus on human-environment relations offers vital insights into how ecologies are being made and remade on private land. By highlighting these dynamics, we offer a fine grain perspective on the topic, which complements the broader institutional and ecological analyses that have examined the need for and the potential roles of private land conservation. Without such a complement, we risk missing vital insights into the social and material worlds in which conservation is performed when considering how to learn, share and adapt conservation goals. The history, scale, speed and diversity of landscape transformations offer further encouragement for grounding our understanding of private land conservation in the contexts in which it occurs. In our case, we focus on the rural-amenity landscapes that surround Melbourne, in Australia, and particularly on conservation practices that involve plants.

In this book, we adopt the term 'rural-amenity landscape' to describe the regions in relatively close proximity to major cities and rural centres, where land is being valued more for the consumption of amenity than it is for agricultural production (Argent, Tonts, Jones, & Holmes, 2010). We refer to those who are taking up residence in these landscapes as rural-amenity migrants. The amenity that these migrants seek is often associated with the 'natural' values and aesthetics of rural areas, the recreational opportunities they provide and the pursuit of 'the simple life' that is perceived to come with leaving the hustle and bustle of the city. Amenity migrants are often as assumed to be retirees (Curry, Koczberski, & Selwood, 2001), but they can also be young families, people buying a second home outside the city (Kondo, Rivera, & Rullman, 2012) or even those willing to commute long distances to city work. As such, the term 'amenity' is deployed here in a broad sense, to capture the diverse aspirations of in-migrants. Despite a history of amenity migration into rural areas stretching back to the 1970s and beyond, its recent acceleration has brought a range of land use and conservation issues to the forefront of policy and academic debate (Abrams, Gill, Gosnell, & Klepeis, 2012; Larsen, Sorenson, McDermott, Long, & Post, 2007).

Rural-amenity landscapes provide an insightful setting in which to explore the finer grain of private land conservation, as they are home to diverse land uses and different land management aspirations. The shifting social dynamics associated with amenity migration means that amenity migrants can have a different perspective on rural landscapes than farmers (Mendham & Curtis, 2010). For example, the pursuit of an 'idyllic' rural lifestyle can manifest in a desire for seclusion on one's land (Meadows, Herbohn, & Emtage, 2013). As a result, amenity migrants can be very 'property-centric' in their ecological interests (Cadieux, 2011). A desire for 'getting on with it' (Gill, Klepeis, & Chisholm, 2010, p. 321) independently on one's own property can contrast with land management in farming communities where collective efforts to address ecological challenges that span property boundaries, such as invasive plants, can be more commonplace.

Amongst the questions associated with rural-amenity migration have been what types of new rural-amenity ecologies will emerge as a result of changing land use (Abrams et al., 2012)? This is especially pertinent in the rural-amenity regions explored in this book, where up to 80% of the pre-colonial vegetation cover has been cleared for agriculture and natural resource extraction (CCMA, 2003). The rural landscape changes associated with amenity migration are often framed as being either positive or negative for ecologies. The negative consequences centre on the potential for species loss and ecological fragmentation as farmland is subdivided into small lifestyle-orientated lots. The preferences of some amenity migrants to pursue conservation for recreational purposes have also been viewed in a negative light, particularly when certain species are retained for aesthetic reasons, as it contrasts with institutional conservation efforts aimed at 'bringing back' a rare local species (Knoot, Schulte, & Rickenbach, 2010; Urquhart & Courtney, 2011; Van Auken, 2010). Running counter to the narrative of negative ecological 'impacts' is the suggestion that rural-amenity migration presents opportunities for catalysing ecological restoration and conservation. A motivation to preserve and restore native ecologies is an aspiration for a host of rural-amenity migrants (Cooke & Lane, 2015; Gill et al., 2010; Mendham, Curtis, & Millar, 2012). Indeed, some landholders purchase rural property with the express intention of pursuing conservation as their primary use activity (Jackson-Smith & Kreuter, 2005; Yung & Belsky, 2007). In recent decades, ecological regeneration has been reported in select rural areas that have seen an increase in amenity migration and a reduction in intensive agriculture (Walker, Marvin, & Fortmann, 2003).

While this book focuses on Australia specifically, similar trends in rural landscape change have been observed in Canada and the United States (Gosnell, 2011), the UK (Phillips, 1993) and across continental Europe (López-i-Gelats, Tàbara, & Bartolomé, 2009; Van Auken, 2010). The shifting socio-ecological dynamics associated with rural landscape change raise significant questions about how new rural landholders will interact with landscapes differently from the production-orientated farmers who preceded them (Gosnell, 2011; Kondo et al., 2012). There is a particular need for analysis of the socio-ecological implications of rural-amenity migration to be grounded in the ecologies and landscape histories in which they play out (Hurley, Maccaroni, & Williams, 2017). The discourse surrounding the ecological consequences of rural-amenity migration also suggests that there is more to examine, and that situated stories of conservation amidst land use change will have much to offer, as we explore the making, remaking and even unravelling ecologies. We now offer some detail as to what it means to think about conservation as a practice that is constituted through human-environment relations.

2 Conservation as a Practice

The notion of conservation practice we articulate here aims to recognise and respond to the divide between nature and culture that still permeates biodiversity conservation discourse. Nature/culture dualism remains the foundation of Eurocentric western scientific thought. It is grounded in the idea that humans have fundamental capacities (namely self-consciousness) that makes them distinct from nonhumans. In broad terms, all living and nonliving things that are not human or human-made belong to the category of nature. Humans and society—knowledge, endeavour and kinship—occupy a domain that is separate, contained and distinct from nature. And not only are nature and culture separate, but ascribing the ability to perceive, reason and contemplate solely to the category of human elevates culture to a status above nature (Watts, 2013). From this standpoint, only humans have agency because acting in the world requires a purpose and intent that only humans possess. Nonhumans are effectively consigned to being a backdrop to the animate world of human society.

What does the idea that only humans possess agency mean for conservation? One of the major implications, and one that we seek to challenge in this book, is the assumption that understanding and interpretation of the nonhuman world develops independently of human encounters with those worlds (Tsing, 2015). From this standpoint, people alone generate knowledge about ecologies, and there is little room to imagine what ecologies could be teaching those who encounter them (Watts, 2013). Further, as the knowledge-holders for conservation, it is people who conduct the practice of conservation, with nonhumans positioned as the objects of conservation action. As has been noted elsewhere, the view that people are the sole arbiters of conservation can promulgate an unjustified sense of human confidence and control over conservation interventions and predictions (Howitt & Suchet-Pearson, 2006). This book is predicated on the idea that, rather than the objects of conservation, nonhumans are *collaborators* in conservation practice. We argue that conservation practice involves the co-production of ecologies through relations and interactions between humans and nonhumans over time.

To give agency to nonhuman entities like plants is to acknowledge that they act in the world and those actions produce meaning that affects the constitution of material things (Todd, 2016). As Latour (2005, p. 71) suggests, agency belongs to 'any *thing* that makes a difference'. What we are suggesting here is that things like plants make an active difference to conservation practice. The ways that 'plants move, assess, evoke, grow, adjust, spread, use, limit and resist' (Head, Atchison, & Phillips, 2014, p. 3), are distinctive capabilities that reveal agency. These capabilities not only shape the way people engage with plants through conservation, but they can dictate what conservation practices are possible through their actions and interactions with other nonhumans. For instance, the ability of a plant species that has been deemed invasive to spread, resist and re-grow can determine whether containment or eradication of the species is the human management objective (Head et al., 2015).

We focus on plants in this book because the kinds of making and remaking of rural-amenity ecologies that we explore frequently revolve around landholders interactions with plants. Indeed, the very notion of re-'vegetation' as a kind of conservation practice reinforces the centrality of plants as focal points of interaction with broader multispecies ecologies. Through planting, tending or removing vegetation, landholders are interacting directly and tangibly with flora. As we will explore, the change, continuity and even death of plants through time can shape human understanding of ecologies in general. In this sense, their interactions with plants can sensitise people to the specific ways that plants work in concert with other nonhuman elements (fauna, invertebrates, soil) to form ecological assemblages on or across private property.

Aside from ascribing nonhuman entities a more active role in our understanding of conservation, nonhuman agency highlights the problems entailed when conservation practices are overly prescriptive or assumed to be easily transferable from one place to the next. When situated interactions are at the core of conservation practice, we can see how such practices generate approaches and ideas about conservation that are distinctly place-based. Moreover, a relational understanding of conservation makes it more attuned to scales and processes that are more-than-human. As we explore in Chapter 2, private land conservation is dictated by the ways people have organised and spatialised the landscape through private property relations. The needs and processes (migration, pollination, food resources) of nonhuman life must conform to the way people organise space from the very outset of the conservation process.

Thinking about conservation practice in less human-centric terms also demands a conscious engagement with a substantive body of First Nations thinking and scholarship on human-environment relations (Hunt, 2014; Ngurra et al., 2019; Todd, 2016; Watts, 2013). Amidst recent interest in nonhuman agency in western scholarship, First Nations scholars have reiterated that Indigenous worldviews—without seeking to generalise Indigenous knowledge—have always considered nonhumans as agentive beings that are part of a society that extends beyond just humans (Watts, 2013). This is reflected in relationships with nonhumans that centre on custodianship rather than the expression of ownership (Graham, 2008), while emphasising that the knowledge that permeates custodianship is situated, nuanced and place-based. In Australia, Indigenous Peoples' relationship to land has been expressed as 'Caring for Country'—a phrase that seeks to challenge colonial logics of command and control over ecologies by recasting land management as care (Suchet-Pearson, Wright, Lloyd, & Burarrwanga, 2013). The reworking of this notion as 'caring-as-Country' takes this further by entangling humans and nonhumans in reciprocal relationships, highlighting that nonhumans also have caring roles (Ngurra et al., 2019; Suchet-Pearson et al., 2013). In this sense, recognising and foregrounding the agency of plants in this book is an acknowledgement of the place of nonhuman agency in First Nations cosmologies, and thus a conscious attempt to challenge colonial framings of conservation as a human-centric endeavour.

Having introduced our notion of conservation practice, let us now return to the experience of Liz and her interactions with a native shrub species on her property called Burgan (*Kunzea Ericiodes*) (see Fig. 1). Burgan is a pioneer species that can rapidly occupy disturbed ground, which gives it a liminal status as a native species that is deliberately removed in parts of West Gippsland due to the belief that is can 'stifle the regeneration process' of other native species (LaTrobe City, 2012). But whether Burgan belongs in landscapes like those in which Liz resides is quite contested. What we see in the way Liz has formed a view on the place of Burgan is an example of the way that the agency of this plant has shaped a trajectory for conservation practice:

> I've been watching [Burgan] for thirty years now. This place was just... a monoculture of Burgan, and so I went for a walk through it one day, and there's birds in it, and there are the spiders and lichen and moss on the ground. And now, thirty years later, it's starting to fall over, and hey, guess

Fig. 1 Burgen (*Kunzea Ericiodes*) is seen by Liz to be a healer of disturbed landscapes. This Burgan patch is beginning to senesce, creating opportunity for other plants to grow

> what, there's all these Eucalypts (endemic tree) in there and I just get so crotchety with people saying Burgan's bad and you should get rid of it. It actually acts like a scab... And the little birds love it, and you gradually get your little [native grasses] and things coming up under it.

It would be easy to gloss over the power of Burgan here, but its capabilities in the form of refuge and habitat for other nonhumans has played a significant role in how Liz views this species. Her conservation practice is enacted *with* Burgan, as they both participate in the slow process of making an ecology that is more complex and nuanced. In this respect, Burgan is seen as a healer, as it protects the landscape in its recovery from past decimation—the 'scab' that it offers is one that eventually falls away to enable new growth and increased species diversity. A process that Liz believes cannot be rushed, which is in contrast to those who seek to remove it in order to speed up regeneration. As refuge and healer, Burgan is shaping knowledge of ecological function and the everyday, hands-on (or in this case, largely hands-off) practice of conservation. This also shows the centrality of plants in understandings of how ecologies act or function in a holistic sense.

We continue to unpack conservation practice through the book as we draw out the myriad ways in which ecologies are being made in rural-amenity landscapes. In particular, Chapter 3 will explore how learning and experimentation processes that inform conservation practice are shaped by nonhuman agency, as rural-amenity migrants become acquainted with the ecologies in which they are immersed. Chapter 4 will further this work as we unpack the connections between conservation practice and the temporality that becomes embodied in landscapes over time. Chapters 5 and 6 then show how thinking about conservation practice as a co-production offers insights into why conservation policies and programs are either co-opted or resisted by both the people and ecologies who participate in them.

3 Introducing the Case Study

The site of this research project was the hinterlands of Melbourne, Victoria, Australia. Two regions were of particular interest, namely the eastern portion of the Corangamite catchment, and the West Gippsland region (see Fig. 2). These rural-amenity landscapes were an appropriate research setting due to the increasing pace of rural-amenity land use transition on the fringes of Melbourne over the last two decades in particular (Mendham & Curtis, 2010), combined with the diverse and patchy ecologies

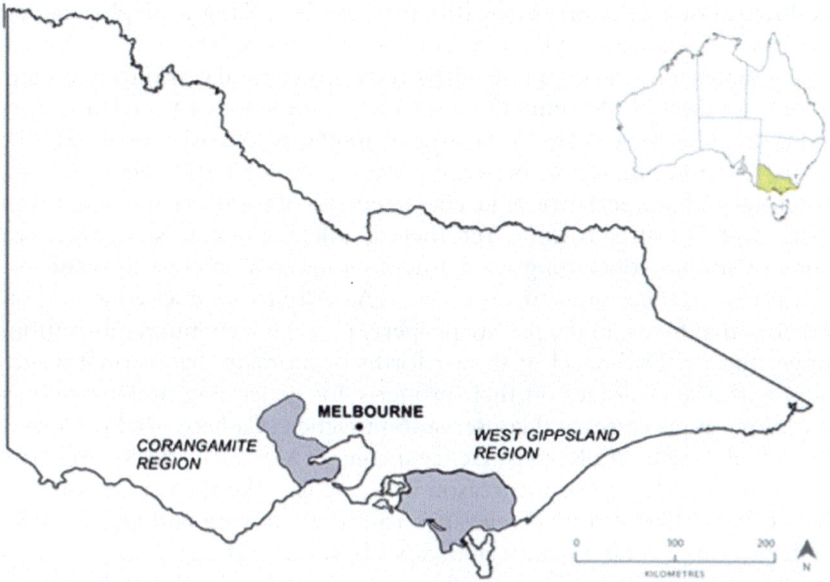

Fig. 2 The location of the rural-amenity regions of Victoria where landholders were visited for the field work

that still persist in these places. In Victoria, coastal and coastal hinterland areas within commuting distance to Melbourne have experienced especially strong amenity migration in this period (Argent et al., 2010). Increasing property prices and land subdivision have also characterised both regions, with land in one of the study areas (East Corangamite) nearly twice the price per hectare of land of similar size and productive potential further to the west of the catchment, suggesting a high amenity premium (Mendham & Curtis, 2010). Price rises and property subdivision are largely indicative of rural-amenity migration (Abrams et al., 2012) and also reflect a growing divide in rural Australian landscapes between areas of high amenity value and areas of low amenity value.

Land use changes away from productive use in select rural regions reflect what could be considered a gentrification of rural landscapes in Australia, with absentee and retiree property ownership on the rise (Race, Luck, & Black, 2010). Yet, this trend of rural gentrification may not be as pronounced as in the United States, where amenity migration is linked with the buy-up of rural land by wealthy individuals seeking to deploy capital they have accumulated in the city (Nelson & Hines, 2018). This characterisation shares much in common with the extensive rural gentrification literature from the UK, detailing the displacement of working rural landscapes and those who work them by wealthy in-migrants (Marsden et al., 2003). Similarities and differences between contexts reinforce the uneven spatial, structural and political economic characteristics of rural-amenity migration across post-industrial nations. Yet, there is a degree of consistency across a range of studies concerning rural-amenity migrants' interest in conservation as part of their move to a rural area. A desire to conduct conservation activities that focus on the private property parcel is a common motivation, suggesting the likelihood of similar forms of intimate, hands-on engagement with the ecologies on their property for generating understandings of ecological processes and conservation practice (Cadieux, 2011; Cooke, 2017; Gill, Chisholm, Klepeis, Denagamage, & Marthick, 2008; Urquhart & Courtney, 2011). For this reason in particular, the themes explored in this book are likely to hold relevance to those studying and interested in conservation in rural-amenity landscapes beyond Australia.

The empirical research that informs much of this book involves three separate phases of fieldwork over a period of nearly a decade. The first was conducted in 2010, involving encounters with 21 participants across both regions identified in Fig. 1. This first phase also saw five interviews with coordinators of private land conservation programs in the two regions, to get their perspective on how policies were being implemented. The second phase of fieldwork was conducted in 2016, involving encounters with 18 landholders, all of whom were from the West Gippsland region. Landholders in the second phase of the fieldwork were all involved in a conservation payment program called 'EcoTender', which forms the basis of the discussion in Chapter 6 (Table 1 in Chapter 3 contains landholder names [pseudonyms], land uses and conservation practices). The third and final phase of the fieldwork is ongoing at the time of writing, involving interviews with past and present bureaucrats, conservation managers, land trusts and environmental non-government organisations (ENGOs), charting the rise and consolidation of private land conservation as a governance approach in Australia. Thirteen interviews have been conducted as part of this project at the time of publication.

The landholders that were visited had resided on their properties for varying lengths of time, ranging from four years to nearly 30 years. This provided a wide time frame over which conservation practices had been conducted and reflected upon. Most participants had moved from suburban Melbourne, with six moving from small properties in rural townships. Seven landholders had some previous farming experience, with two having retired on the property they previously farmed. More than two-thirds of the collective cohort participated in some form of conservation program on their property, which highlighted a strong interest in conservation overall. Participation in private land conservation programs like Trust for Nature conservation covenants (see Chapter 5) was used as a recruitment strategy for the research. Landholders with remnant vegetation on their property (primarily identified through aerial photography) or properties where there was visible evidence of ecological restoration being conducted, had an invitation to participate in the research dropped in their letter box.

The encounters with landholders in 2010 and 2016 involved two primary research methodologies—semi-structured interviews in the home and a walking interview around the property. The initial interview in the home proved to be incisive for understanding how landholders thought about their conservation practices, what motivated them to manage their properties in the manner that they did, and for charting the change and continuity in their local landscapes over time. Where possible, participants were invited to share photos from the early phases of their property tenure, to help ground some of the discussion of conservation practice and landscape change. Benjamin (first author) also provided aerial photos of their properties that included local landscapes where they were available, to help situate discussion of landscape-scale processes or the action (or inaction) of neighbours as they related to conservation.

A second interview was then conducted in a more active and informal manner by venturing outside and continuing the discussion while walking around the property. This provided critical insights for understanding how people's perspectives on conservation were playing out through their activities. This approach recognised that the material environments that are of importance to people embody memories and experiences of interaction with those spaces. Allowing the landscape to serve as a repository of memory for landscape change and conservation practice also offered an insight into the ways that ecologies themselves were shaping conservation practice. In this sense, the walking methodology attributed some agency to the landscape as a participant in the research process. Observing the evidence of ecological change over time (remnant vegetation or past efforts at revegetation, for example), the varying growth patterns and fortunes of different plants, competition between different species for resources and evidence of plant regeneration and spread over time was critical to the themes explored here. Walking interviews also meant that observations of plants in situ could then be compared to landholders' descriptions of how their practices and perspectives had been shaped by encounters with plants inhabiting their property. Observations of the physical outcomes of management practice also provided evidence of how management practices had shifted over time. These changes could reveal themselves in terms of different planting strategies over time or even the selection of different species for planting.

The walking methodology also brought a sense of dynamism to the research process, as we traversed tracks from one space of conservation activity to the next. For instance, participants noted the presence of 'weedy' plants as we walked the property, with five participants gravitating towards specific property spaces to inspect the weeds and remove them by hand—a process in which Benjamin (first author) also participated. In some cases, these events would redirect our movements away from a particular path as we searched for more weeds to remove, which also prompted discussion about the challenges of 'keeping on top of' your conservation work. Photos were also taken as we walked around, and some of these images feature in the following chapters.

4 Book Structure and Contents

Having established a working definition of conservation practice here, Chapter 2 begins with a foundational examination of western, liberal private property relations. In this chapter, we unpack exactly what we mean when we are talking about private property, with specific attention to the fact that property relations are contested. This is important to understand, as the influence of property is often unappreciated in conservation discussions. We characterise western liberal property as fundamentally a social relation between people, with respect to land. With this in mind, we directly address how private property influences the way ecologies are being made (or not) on private land, the misalignment between property as a form of spatial organisation for humans and the scales and processes of nonhuman life forms, the need for more emphasis on the obligations rather than the rights entailed in property ownership, and the extent to which property rights may need to be re-imagined in light of social-ecological change.

In Chapter 3, we explore how conservation practice is informed through the everyday interactions between people and the ecologies they inhabit.

A number of the rural-amenity migrants had no prior familiarity with their local landscapes, so we examined how they formed strong dispositions for conservation through learning-by-doing. We show that the response from local ecologies to their interactions and interventions is pivotal in determining whether landholders take a more active and interventionist approach to conservation, or a more passive and hands-off approach. This chapter makes the case for bringing nonhumans, particularly plants, more formally into our understanding of experiential learning.

In Chapter 4, we examine how historical landscape modification combines with ongoing environmental change to shape conservation on private land. We propose the notion of 'landscape legacy' to show how rural-amenity migrants are inheriting a history of conservation practice and land use that is embodied in the landscape. Through clearing for colonial agriculture, fencing, farming and historic restoration attempts, landholders' efforts to create their own legacy of ecological enhancement through conservation are mediated by past legacies embodied in the landscape. In some cases, these past legacies serve as a trajectory for the future, as current landholders carry forward practices or promote species whose capabilities resonate with them.

In Chapter 5, we look at the adoption of conservation covenants on private land. Having demonstrated how human-environment interactions co-constitute private land conservation practice, we examine how policies and programs aimed at private land conservation are mediated through those practices. In the case of covenants, we build from the previous chapter on landscape legacy to show how landholders use covenants as a permanent protection mechanism, not just to secure ecological values on private property, but also to secure their legacy of ecological care and labour over time. We also show that covenants as a legal instrument do not guarantee the continuity of conservation practices and explore how landholders grapple with this realisation.

In Chapter 6, the focus shifts to the use of market-based instruments for delivering conservation outcomes on private land. We examine the case of a reverse-auction tender mechanism called 'EcoTender', placing particular emphasis on the way that the market logic of the scheme fails to align with how conservation practices are conducted or conceived. We specifically focus on how landholders approach the costing of their own labour as part of the bidding process for competitive tenders, the ways in which ecologies may resist program prescriptions framed around discrete private property parcels and explore the contradictions in participants' desire for social interaction with fellow participants in a program that requires competition to drive down costs. We conclude by discussing the role of market-based instruments (MBIs) amidst local socio-ecological contexts and environmental change.

We conclude the book with reflections on the need for more collective approaches to conservation in rural-amenity landscapes, and the central role that practice-sharing needs to play in this process. A collective conservation focus is then tied into a discussion of the need for new goals to be articulated for private land conservation in which we emphasise the importance of collaborating around place-based conceptions of ecologies. We then consider whether more fundamental change is needed in how private property rights are conceived and enacted to fully realise collaborative conservation in ways that acknowledge nonhuman agency, as well as challenging settler-colonial approaches to private land conservation that take place on the unceded lands of First Nations Peoples.

References

Abrams, J., Gill, N., Gosnell, H., & Klepeis, P. (2012). Re-creating the rural, reconstructing nature: An international literature review of the environmental implications of amenity migration. *Conservation and Society, 10*(3), 270. https://doi.org/10.4103/0972-4923.101837.

Argent, N., Tonts, M., Jones, R., & Holmes, J. (2010). *Demographic change in Australia's rural landscapes* (G. W. Luck, R. Black, & D. Race, Eds., pp. 23–44). Landscape Series, 12. Dordrecht: Springer Netherlands. https://doi.org/10.1007/978-90-481-9654-8.

Cadieux, K. V. (2011, July). Competing discourses of nature in exurbia. *GeoJournal, 76*, 341–363. https://doi.org/10.1007/s10708-009-9299-0.

CCMA (Corangamite Catchment Management Authority). (2003). *Corangamite regional catchment strategy 2003–2008*. Colac: Corangamite Catchment Management Authority.

Cooke, B. (2017). The co-presence of past and future in the practice of environmental management: Implications for rural-amenity landscapes. In *Nature, temporality and environmental management: Scandinavian and Australian perspectives on peoples and landscapes* (pp. 77–93). London: Routledge.

Cooke, B., & Lane, R. (2015). How do amenity Migrants learn to be environmental stewards of rural landscapes? *Landscape and Urban Planning, 134*, 43–52. https://doi.org/10.1016/j.landurbplan.2014.10.006.

Curry, G. N., Koczberski, G., & Selwood. J. (2001). Cashing out, cashing in: Rural change on the south coast of Western Australia. *Australian Geographer, 32*(1), 109–124. https://doi.org/10.1080/00049180020036268.

Gill, N., Chisholm, L., Klepeis, P., Denagamage, R., & Marthick, J. (2008). *Land management and land cover on land owned by amenity oriented rural landowners in Jamberoo Valley*. https://ro.uow.edu.au/cgi/viewcontent.cgi?referer=&httpsredir=1&article=2039&context=scipapers.

Gill, N., Klepeis, P., & Chisholm, L. (2010). Stewardship among lifestyle oriented rural landowners. *Journal of Environmental Planning and Management, 53*(3), 317–334. https://doi.org/10.1080/09640561003612890.

Gosnell, H. (2011). Amenity migration: Diverse conceptualizations of drivers, socioeconomic dimensions, and emerging challenges. *GeoJournal, 76*(4), 303–322. https://doi.org/10.1007/s10708-009-9295-4.

Graham, M. (2008). Some thoughts about the philosophical underpinnings of Aboriginal worldviews. *Worldviews: Global Religions, Culture, and Ecology, 45*, 1–9.

Hardy, M. J., Fitzsimons, J. A., Bekessy, S. A., & Gordon, A. (2017). Exploring the permanence of conservation covenants. *Conservation Letters, 10*(2), 221–230. https://doi.org/10.1111/conl.12243.

Head, L., Atchison, J., & Phillips, C. (2014). The distinctive capacities of plants: Re-thinking difference via invasive species. *Transactions of the Institute of British Geographers, 40*(3), 399–413. https://doi.org/10.1111/tran.12077.

Head, L., Larson, B., Hobbs, R., Atchison, J., Gill, N., Kull, C., & Rangan, H. (2015). Living with invasive plants in the Anthropocene: The importance of understanding practice and experience. *Conservation and Society, 13*, 311–318.

Howitt, R., & Suchet-Pearson, S. (2006). Rethinking the building blocks: Ontological pluralism and the idea of 'management'. *Geografiska Annaler, Series B: Human Geography, 88*(3), 323–335. https://doi.org/10.1111/j.1468-0459.2006.00225.x.

Hunt, S. (2014). Ontologies of indigeneity: The politics of embodying a concept. *Cultural Geographies, 21*(1), 27–32. https://doi.org/10.1177/1474474013500226.

Hurley, P. T., Maccaroni, M., & Williams, A. (2017). Resistant actors, resistant landscapes? A historical political ecology of a forested conservation object in exurban southeastern Pennsylvania. *Landscape Research, 42*(3), 291–306. https://doi.org/10.1080/01426397.2016.1267131.

Jackson-Smith, D., & Kreuter, U. (2005). Understanding the multidimensionality of property rights orientations: Evidence from Utah and Texas ranchers. *Society and Natural Resources, 18*, 587–610. https://doi.org/10.1080/08941920590959578.

Kay, K. (2016). Breaking the bundle of rights: Conservation easements and the legal geographies of individuating nature. *Environment and Planning A, 48*(3), 504–522. https://doi.org/10.1177/0308518X15609318.

Knoot, T. G., Schulte, L. A., & Rickenbach, M. (2010). Oak conservation and restoration on private forestlands: Negotiating a social-ecological landscape. *Environmental Management, 45*(1), 155–164. https://doi.org/10.1007/s00267-009-9404-7.

Kondo, M. C., Rivera, R., & Rullman, S. (2012). Protecting the idyll but not the environment: Second homes, amenity migration and rural exclusion in Washington State. *Landscape and Urban Planning, 106*(2), 174–182. https://doi.org/10.1016/j.landurbplan.2012.03.003.

Larsen, S., Sorenson, C., McDermott, D., Long, J., & Post, C. (2007). Place perception and social interaction on an exurban landscape in Central Colorado. *The Professional Geographer, 59*(4), 421–433. https://doi.org/10.1111/j.1467-9272.2007.00632.x.

Latour, B. (2005). *Reassembling the social—An introduction to actor-network-theory*. Oxford: Oxford University Press.

LaTrobe City. (2012). *Amendment C102—Removal of Burgan and public acquisition overlay correction*. http://www.latrobe.vic.gov.au/Building_and_Planning/Development/Planning_Scheme_Amendments/Completed_Planning_Scheme_Amendments/Amendment_C102_%E2%80%93_Removal_of_Burgan_and_Public_Acquisition_Overlay_Correction. Accessed 7 May 2019.

López-i-Gelats, F., Tàbara, J. D. & Bartolomé, J. (2009). The rural in dispute: Discourses of rurality in the Pyrenees. *Geoforum, 40*(4), 602–612. https://doi.org/10.1016/j.geoforum.2009.04.008.

Marsden, T., Murdoch, J., Lowe, P., & Ward, N. (2003). *The differentiated countryside*. London: Routledge.

Meadows, J., Herbohn, J., & Emtage, N. (2013). Supporting cooperative forest management among small-acreage lifestyle landowners in Southeast Queensland, Australia. *Society & Natural Resources, 26*(7), 745–761. https://doi.org/10.1080/08941920.2012.719586.

Mendham, E., & Curtis, A. (2010). Taking over the reins: Trends and impacts of changes in rural property ownership. *Society & Natural Resources, 23*(7), 653–668. https://doi.org/10.1080/08941920801998893.

Mendham, E., Curtis, A., & Millar, J. (2012). The natural resource management implications of rural property turnover. *Ecology and Society, 17*(4), 5. https://doi.org/10.5751/ES-05071-170405.

Nelson, P. B., & Hines, J. D. (2018). Rural gentrification and networks of capital accumulation—A case study of Jackson, Wyoming. *Environment and Planning A, 50*(7), 1473–1495. https://doi.org/10.1177/0308518X18778595.

Ngurra, D., Dadd, L., Glass, P., Scott, R., Graham, M., … Suchet-Pearson, S. (2019). *Yanama budyari gumada*: Reframing the urban to care as Darug Country in western Sydney. *Australian Geographer, 50*, 279–293. https://doi.org/10.1080/00049182.2019.1601150.

Phillips, M. (1993). Rural gentrification and the processes of class colonisation. *Journal of Rural Studies, 9*(2), 123–140. http://www.sciencedirect.com/science/article/pii/074301679390026G.

Race, D., Luck, G. W., & Black, R. (2010). Patterns, drivers and implications of demographic change in rural landscapes. In G. Luck, R. Black, & D. Race (Eds.), *Demographic change in rural landscapes: What does it mean for society and the environment?* (pp. 1–22). The Netherlands: Springer.

Suchet-Pearson, S., Wright, S., Lloyd, K., & Burarrwanga, L. (2013). Caring as country: Towards an ontology of co-becoming in natural resource management. *Asia Pacific Viewpoint, 54*(2), 185–197. https://doi.org/10.1111/apv.12018.

Todd, Z. (2016). An indigenous feminist's take on the ontological turn: 'Ontology' is just another word for colonialism. *Journal of Historical Sociology, 29*(1), 4–22. https://doi.org/10.1111/johs.12124.

Tsing, A. L. (2015). *The mushroom at the end of the world: On the possibility of life in capitalist ruins.* Princeton, NJ: Princeton University Press.

Urquhart, J., & Courtney, P. (2011). Seeing the owner behind the trees: A typology of small-scale private woodland owners in England. *Forest Policy and Economics, 13*(7), 535–544. https://doi.org/10.1016/j.forpol.2011.05.010.

Van Auken, P. M. (2010). Seeing, not participating: Viewscape fetishism in American and Norwegian rural amenity areas. *Human Ecology, 38*(4), 521–537. https://doi.org/10.1007/s10745-010-9323-5.

Walker, P. A., Marvin, S. J., & Fortmann, L. P. (2003, December). Landscape changes in Nevada County reflect social and ecological transitions. *Environmental Toxicology & Chemistry, 57*, 115–121.

Watts, V. (2013). Indigenous place-thought & agency amongst humans and non-humans (First Woman and Sky Woman go on a European world tour). *Decolonization: Indigeneity, Education & Society, 2*(1), 20–34.

Yung, L., & Belsky, J. (2007). Private property rights and community goods: Negotiating landowner cooperation amid changing ownership on the Rocky Mountain Front. *Society & Natural Resources, 20*(8), 689–703. https://doi.org/10.1080/08941920701216586.

CHAPTER 2

Private Property Relations and Conservation Practice

Abstract Private property is a central component of conservation on private land, but what exactly is meant by private property and how does it shape the potential for conservation practices? This chapter unpacks western, liberal private property before examining its conservation implications. As private property is deeply implicated in the spatial and temporal configuration of landscapes, it has substantial consequences for how conservation is understood and enacted. As we will outline, we need to think less passively about nature conservation as happening *on* private land, to an awareness that it happens *through* an interplay of property, people and ecologies. This awareness allows us to see whether and how configurations of property rights need to shift in order to meet conservation challenges amidst socio-ecological crisis.

Keywords Private property · Ownership · Property rights · Conservation · Nonhuman

1 Introduction

Private property is a central component of conservation on private land, but what exactly is meant by private property? What forms does it take and how does this shape the potential for conservation practices? To answer

B. Cooke and R. Lane, *Making Ecologies on Private Land*,
https://doi.org/10.1007/978-3-030-31218-3_2

these questions, we begin by explaining what we mean by 'private property' before examining its broader implications and importance for understanding nature conservation practice and policy. As Keenan (2014) notes, liberal western conceptions of property have positioned land as 'blank and inert' (p. 25), which can mask the way that property actively contributes to the production of particular types of landscapes. Passive and apolitical representations of private property can then allow narratives of absolute ownership to progress unchallenged and unexamined. As we will outline here and further examine in subsequent chapters, property is far from a neutral foundation on which the more pressing or challenging dimensions of conservation are layered.

This book is motivated by the need for more critical reflection around the influence of private property in conservation policy and practice, and the need to recognise property as a rigid but also historically changeable element of the human-environment relations that shape conservation practice. Private property is deeply implicated in the spatial and temporal configuration of landscapes, which means it has substantial consequences for how conservation is understood and enacted. And it is only by engaging with the influence of property that we can contemplate possible re-imaginings of property in aid of conservation. As we will outline, we need to think less passively about nature conservation as happening *on* private land, to an awareness that it happens *through* an interplay of property, people, institutions and ecologies. This awareness allows us to see whether and how configurations of property might need to shift in order to meet the challenges of conservation on private land.

In this chapter, we articulate a definition of private property as a 'relation' that is concerned with land, in a western liberal context. The emphasis on relations is deliberate and important, as it highlights that property ownership and rights are not actually concerned with the relationship between people and a specific parcel of land. Private property is not a material object or 'thing' in this instance. Rather, private property is understood as a social relationship between people *with respect* to land, which bestows certain rights, power, privilege and responsibilities to the landholder (Keenan, 2014).

Following our introduction to private property, we then introduce the avenues by which property relations can shape ecologies and conservation practices in rural-amenity landscapes. We will build on this discussion in later chapters through our focus on environmental stewardship dispositions, motivations for permanent protection of ecologies and the use of market instruments in conservation. The way private property weaves through the stories of rural-amenity conservation that we encounter later on will demonstrate its role in the practices of conservation on private land. As such, this chapter sets the scene for how private property not only fragments rural-amenity ecologies, but the way it also produces new and novel ecologies.

The examples of private land conservation activities explored in this book are drawn from Australia. However, the work that property does to shape ecologies in the landscape, and the way it shapes social interactions with landscape, will resonate with a wider readership. This is especially true of private land conservation occurring in other settler-colonial states. But as the literature on conservation and landscapes is beginning to show, key issues relating to sense of belonging, the right to exclude, land use change and social learning, will often intersect with private property relations, however represented or enacted (Brown, Flemsæter, & Rønningen, 2019; Cooke & Lane, 2018; Ojalammi & Blomley, 2015; Yung & Belsky, 2007). Property relations are bound up with a host of conservation practices and policies—not just on private land, but public land too.

2 Private Property Relations in Land: The Western Liberal Tradition

It is worth noting from the outset that there is no singular expression of private property, even in the context of western liberal democracies. There is a dominant form of property that is more likely to be recognised in law and through planning systems, but this is not settled and is frequently up for negotiation. In a similar vein, there is no universal landholder. We use the term landholder here specially to acknowledge that 'ownership' of the land may be contested. In Australia, where the Federal Government has no formal treaty with First Nations, who never ceded sovereignty of the land, we use the term 'landholder' to acknowledge that 'ownership' is contested. While this book focuses largely on individual landholders, landholders can come in many forms when it comes to private land conservation. Private land trusts, environmental non-government organisations (ENGOs), corporate entities, recreational groups and community collectives are all examples of landholders that could be pursuing conservation on private property.

When we talk of private property relations, we are referring specifically to 'real property', or private property held in the form of land. We are also talking about property in the western liberal tradition that has become the dominant expression of property in Australia since colonial invasion. As noted above, we wish to reinforce the idea that property relations define the way people relate to one another concerning land, as this negates the notion of land ownership as an absolute right possessed by a landholder. While Australia shares a great deal of commonality with other western liberal states in terms of the expression of private property rights in legal, planning and vernacular terms, private property has different expressions in different places. For example, the existence of 'right to roam' laws in the UK, which can provide public access to designated private land for recreational purposes, is not something that exists in Australia. Despite these variations, there are core tenets of private property relations shared across many liberal property systems that are relevant to a broader conservation community.

Owning real property, from this relational understanding of property, means possessing a set of rights, with the most fundamental of these rights being the ability to exclude non-rights-holders from sharing in certain benefits of that property (Blomley, 2016). Indeed, real property in all forms (public, private, communal) bears some association with the ability to exclude others. As an exclusive rights-holder, a private property owner can then benefit from usage rights in a given space (including things like the right to exploit land for primary production). The kinds of activities one may conduct on private property, and who one may exclude, is mediated by the legal and planning instruments of a given jurisdiction. But in general terms, the nonhuman elements, both living and inert, that are present on land over which private property rights are held, are subsumed into that property through those rights.

Thinking about property as a set of rights governing social relations is often referred to in legal terms as a 'bundle of rights' (Mansfield, 2008). In effect, private property ownership is not a singular right, but a combination of more specific rights to the relevant land that, when bundled up, constitute private land ownership. As an example, in Victoria, Australia, the Mines Resource (Sustainable Development) Act 1990 declares that all minerals in the earth's crust belong to the Crown (State Government). This diverges from the common law precedent that all minerals (surface and substrate) belong to the private property holder (Hepburn, 2011). It is this law that means that a minerals exploration licence can be granted to an individual or corporation that is not the private landholder. This situation has been usefully described as 'topsoil rights', where the owner of the land title only owns the land surface (Adams & Moon, 2013). A rights-holder in land may also be restricted in pursuit of other rights through legal or regulatory mechanisms. The right to remove a plant or animal species listed as endangered on private land may not be allowed, at least not without applying for a permit through the relevant statutory authority. The bundle of rights analogy helps to remind us that private land ownership does not grant absolute and unrestricted rights over a designated area, and that private property relations are governed by legal and regulatory instruments that are indivisible from general notions of private ownership.

While some popular discourse on private property rights suggests a sanctuary from the interference of government in the affairs of the individual, in reality, private property is deeply tied to the activities of the state. The rights that are granted through private property ownership are backed by the power of the state, and the threat of violence or prosecution for transgression of those rights by non-rights-holders (Keenan, 2014). Without the threat of enforcement, the legitimacy of private property rights is diminished. In western liberal contexts today, the enforcement of property rights often hinges on the registration of land title in the jurisdiction where the land is located. Each property has a single title, and the possessor of that land is the person or entity that is reflected in the relevant register of titles. Title registration began with the Torrens title system in 1850s colonial Australia, but has now spread rapidly around the globe. Rather than having to compile a record of demonstrated ownership through occupation and land use over time (up to 60 years), as was the case with common law property systems, land titles solidify property as 'a priori, singular, and definite' (Blomley, 2014, p. 1291). Amongst other things, title registries enable a simplified process for land purchase and sale, in a fashion that is equivalent to most commodities.

As a technology that contributes to our understanding of private property, titles are also accompanied by cadastral representations of property through maps. Mapping is a vital technology of property as it plots out the boundaries, location and size of a parcel of land. Both titles and maps are enfolded into the governance processes of the state and help to construct a particular understanding of private property as a claim to a bounded parcel of land (Blomley, 2016; Li, 2014). Cadastral maps play a notable role in the representation of property as a unit of land that is easily extractable—an object that can be divorced from its landscape context and the particularities of the place in which it is situated (Graham, 2011). Indeed, one of the consequences of cadastral mapping is the production of parcelised representations of private property that can render invisible more nuanced understandings of property. Interpretations of private property as relational and a bundle of rights are obscured by the way property as land is represented and subsequently enacted in the landscape.

The way that property materialises in the landscape is often the way we understand and experience it from the standpoint of conservation practices. In emerging areas of legal geography, the everyday enactment of property is articulated as a kind of 'territory' that is created through property ownership. As Blomley (2016) notes, the property territory is represented as a uniform material space in which the rights to use, access or to exclude others are performed. This territory is denoted by the property boundary (usually in the form of a fence line), which marks out the space of territorial claim. In this way, territory and property are co-produced as rights materialise spatially, and the experience of enclosed property space informs an understanding of property. That we enact private property ownership in western liberal contexts as a territorial claim over a distinct parcel of land may seem self-evident. However, the implications of the way that property organises space for conservation are rarely discussed explicitly, despite the fact that it 'structures power relations in specific and highly consequential ways' (Blomley, 2016, p. 595). When it comes to conservation, we argue that our understanding and analysis of policy and practice often fail to reflect consciously and carefully on the structural role of property. Not only does this limit our ability to interpret the influence of property in conservation challenges on private land, but it cedes all power of how private property should be interpreted and represented to those wishing to assume absolute rights over land with minimal responsibilities to non-rights-holders.

While property rights are understood in law as a bundle of non-exclusive rights, popular rhetoric around what private property entails is often more absolute. Tensions between legal and popular interpretations of property have frequently manifested in public debate over tree clearing laws in Australia and elsewhere. For example, a former Australian Agriculture Minister, Barnaby Joyce, once labelled regulations restricting tree clearing as a means of 'dispossess(ing) people of their private assets' (ABC News, 2016, October 26). Through this lens, everything present within the owned parcel of land is a resource for private exploitation. This attitude—which is far from unique—positions planning and legal restrictions as an unjust restraint on the full expression of private property rights. Mansfield (2008) has suggested that such a position is bound up with a broader imperative for maximising the commodity value of land through the rhetoric of privatisation as an inalienable right. To be useful as a means of capital accumulation she argues, property ownership must enable "complete control... relatively unfettered by broader social restriction" (Mansfield, 2008, p. 7). From this perspective, property as a relation is something that only bestows rights; the idea that property ownership might also bestow responsibilities is removed from the equation.

The framing of property rights is frequently communicated in terms of individual control. The notion of individual control of a property and its management appears to be particularly strong in the kinds of rural-amenity landscapes we focus on here. Not only has rural-amenity migration seen further enclosure of rural lands through property subdivision and altered land use regimes (Mendham & Curtis, 2010), in-migrants themselves can see the property parcel as an insulated refuge from a wider socio-ecological context. Even when it comes to conservation practice, the motives for pursuing it are often entangled with a desire to improve the aesthetic appeal of the landscape and to 'experience nature as an individual' (Cadieux, 2011, p. 349). The conservation activities that landholders pursue can be linked with a desire to undertake conservation within the 'controlled' space of the property parcel, rather than the unknowns of public or communal land (as we touch on in Chapter 3). In our own research, we have found that rural-amenity migrants can possess an attitude to private property that is more absolute than primary producers. Other researchers have also provided evidence of rural-amenity migrants shunning established social norms around when private property may be accessed by others—for example when livestock have broken through a fence (Yung & Belsky, 2007).

What is clearly missing from absolutist interpretations of private property ownership is the notion that ownership also carries responsibilities (Graham, 2011). Against this skewed interpretation, we argue that the obligations on the rights-holder must be more clearly articulated. An obligation that is easy to ignore amidst the imaginaries of parcelised and commodified land that is abstracted from its environmental and societal context and managed as an isolated entity. As Gammage (2012, p. 321) notes, 'fences on the ground make fences in the mind'. One of the challenges for articulating property obligations is the need to orientate them towards neighbours, the broader community and to nonhumans and the landscape (Cooke & Lane, 2018). In order to chart a way forward for a more relational understanding of private property, it may be necessary to lift our gaze beyond the property space. As Graham (2011, p. 206) notes, an articulation of property law that better recognises our obligations to the nonhuman should look no further than the landscape, as the 'landscape reveals… our place in the world and the opportunities and limits of our connection with it'.

3 Private Property and the Structuring of Conservation Practice

One of the dangers for private land conservation is that the dominant relationship underpinning conservation practice becomes that of property owner and property territory, which renders invisible the complex interactions of human and nonhuman elements within dynamic landscapes. In this sense, property as a parcel of territory can obscure the fact that land has a 'natural attitude' (Linebaugh, 2008, p. 45). Species interactions and biophysical processes rarely align with the boundary of private property. While fence lines may intentionally denote the property parcel as private and exclusive space to other humans, private property and its manifestations in the landscape clearly impact on nonhuman lives too. The fragmenting of vegetation communities and disruption to species migration are common symptoms of a parcelised landscape (Jakes, Jones, Paige, Seidler, & Huijser, 2018). This parcelisation and the creation of enclosed territories through the enforcement of boundaries pose challenges for the increasing calls for a landscape or ecosystem-scale approach to conservation in ecology and for more dispersed and contingent human-environment relations in political ecology (Hodgetts, 2018).

While private property can pose fragmentation risks for ecologies, in this book we are most interested in how it can shape conservation practices in the context of rural-amenity landscapes. We see private property relations as having a role in the making of new and novel socio-ecological spaces. Some of the influences of property relations on conservation practice are easily observable through hedge and fence line plantings that entrain ecologies to the boundaries of property. They might also include the use of plants as screening for privacy from a neighbour, with consequences not just for the ecological assemblage, but for shaping social relationships between neighbours. Other influences are likely to be more subtle, relating to how notions of rights and responsibilities associated with land ownership lead to the pursuit of certain conservation practices and the absence of others. In the chapters that follow, we show how property relations become entwined in conservation and the particular kinds of ecologies this produces; how notions of rights over land become internalised and are enacted in the practice of conservation. This can be true even for committed conservationists, who may not acknowledge the role of property in shaping their relationship to ecologies. Many rural-amenity migrants are reticent to tell their neighbouring landholders what to do with 'their land', even if those suggestions involve substantially improved conservation outcomes (Cooke & Lane, 2015). As a result, there is the potential for a very 'property-centric' approach to conservation in rural-amenity landscapes, with each property defining a different management regime.

Rural-amenity landscapes present an amplified setting in which to explore the influence of private property relations on conservation because of their parcelled character. The challenge of undertaking collective action tasks like invasive animal control across a landscape of numerous small properties is a classic example of such an issue (Marshall, Coleman, Sindel, Reeve, & Berney, 2016), especially when the purchase of rural-amenity land has not been accompanied by a sense of responsibility for controlling weeds for the sake of neighbours. The spatial scale of property enclosed within boundary fences is clearly incompatible with the mobility of plant species (Fiege, 2005), whether desired or undesired. Returning to our call for more attention to nonhuman agency in the opening chapter, we can see how existing private property relations do not reflect the lives, needs and capabilities of nonhumans, and how this will need to change (Brown et al., 2019).

Despite the issues around private property outlined here, it should be noted that the influence of a property-centric mentality for conservation is far from consistent, nor is it uniformly problematic. Moreover, property tenure in isolation does not determine conservation outcomes on private land. Geographers have highlighted that a range of ownership rationalities can be enacted within any one category of land tenure (Massey, 1980). For example, the idea that property ownership is more of a custodial role than an exercising of rights is not foreign to landholders interested in conservation, with one landholder noting:

> at the end we're only custodians of the land. You don't own it. You own it on paper, but you're not going to last forever, so you're passing it on to somebody. (Adam)

Adam's view here suggests a kind of temporal obligation for caring for the ecologies on his property beyond his tenure. Even more exclusive attitudes to land ownership may also have a conservation benefit. One example of this is the way that property rights and fences that exclude others can be the only thing that keeps introduced predators away from threatened species. The popularity of predator-proof fencing to protect threatened species in discrete pockets of the landscape is proving increasingly popular as a conservation strategy in Australia. The isolation that comes from applying abstract ideas of property in practice may offer a kind of refuge amidst landscapes heavily modified by colonial landscape change.

When we consider the future challenges that face conservation practice on private land, property-centric approaches must be a central concern. The combination of more rapid and expanded species migration due to climate change and the increasing compartmentalisation of rural landscapes through property subdivision (especially in rural-amenity landscapes) suggest that property boundaries and private property rights will be a flashpoint for conservation and land management conflicts (Pecl et al., 2017). Just as fences and walls on sovereign borders are increasingly impacting vulnerable and displaced peoples, fences in rural and peri-urban landscapes have and will continue to have substantial consequences for nonhuman species on the move. Ecosystem-scale conceptions of ecologies are needed in order to respond to species mobility challenge, but that requires greater cooperative efforts across landscapes. Will meeting the unfolding ecological crisis require a re-imagining of property in pursuit of collective, landscape-scale conservation practice?

In the following chapters of this book, we engage with the specific and nuanced influence of private property rights on conservation practices. Chapter 3 examines how pursuing conservation in a space that is yours, over which you have control, is embedded in experiential learning and environmental stewardship. Property rights surface as both a driver of conservation action and a source of concern when thoughts turn to the future of conservation efforts. Notions of being a steward of the land and caring for nature are inevitably bound up with ideas of ownership. This discussion shows that when land is treated as a thing that is owned rather than a place that has character and particularities undefined by property boundaries, there is a risk of entraining conservation practice to property rather than place (Graham, 2011). Yet, as we show, for some landholders, the right to pursue conservation 'on my land' can be intimately tied up with a sense of responsibility to care for ecologies.

In Chapters 5 and 6, we show how conservation policy and governance processes need to do more to account for the implications of private property in their design and implementation. The expectation that there is a baseline level of conservation action that landholders should undertake as part of an ownership responsibility is an active point of discussion. There are many conservation programs in operation that offer payments for conservation work, or provide permanent legal protection for ecologies through covenants (easements), that are grounded in the idea of property as a spatially discrete parcel. However, when covenants are used to help preserve ecologies, they are actively challenging absolute property by ceding particular use and exploitation rights associated with private property. As we explore though, the extent to which this legal mechanism supports ongoing conservation in the case of property turnover remains uncertain for some landholders. These dynamics of property relations have a clear intersection with how conservation practice manifests in material ways.

4 Conclusion

By positioning conservation practice as bound up in property, we wish to raise the prospect that policy and advocacy work that seeks to shift conservation practices must consider the implications for and extending from property relations. We offer this chapter as a primer to the empirical work that we discuss in subsequent chapters to set the scene for understanding how private property relations are dynamic and contested. It should be remembered that private property is understood and expressed in a host of different ways that complicate any attempt at neat categorisation, and there is always a performative dimension to how property is enacted in practice, which can blur easy definitions of private property. While our work focuses on individual landholders, the questions we raise about thinking through private property relations are also relevant for the use of private land by land trusts, ENGOs and other stakeholders in the conservation sector. In the final chapter of the book, we consider the need for conservation practice to operate more collectively across property boundaries in rural-amenity landscapes, as well possibilities for how private property relations could be organised differently, especially in responding to nonhuman agency and the challenge of addressing the settler-colonial foundations for conservation in places like Australia.

The role and influence of private property rights must not be ceded to a neoliberal discourse of ownership that has no interest or concern with responsibility and obligation. Not only does this blind us to the influence of property rights and landholder perceptions of ownership of ecologies, but it limits our ability to conceive of new possibilities for private property.

These new possibilities may be as simple as stronger articulations of ownership obligations, or as radical as a fundamental rethinking of property rights law so that it is guided by the needs of the landscape (Graham, 2011). As has also been highlighted in the context of conservation (Cooke & Lane, 2018) and in housing (Crabtree, 2017), the displacement and mobility of humans and nonhumans that is accompanying socio-ecological changes may not be compatible with our current enactment of private property rights. It may be that a rethinking of property rights is not just a means of dealing with socio-ecological challenges, but a necessity driven by global environmental change. Whatever the case, we need rich and qualitative inquiry that explores how private property rights are shaping the everyday practices of conservation in order to inform such discussions. Rural-amenity landscapes provide an excellent context in which to do this, given the heightened socio-ecological fragmentation that is characteristic of these places.

References

ABC News. (2016, October 26). *Joyce accuses ALP of adopting communist-style land clearing laws*. https://www.abc.net.au/news/2016-10-26/joyce-accuses-alp-of-adopting-communist-style-land-clearing-laws/7967932. Accessed 14 February 2019.

Adams, V. M., & Moon, K. (2013). Security and equity of conservation covenants: Contradictions of private protected area policies in Australia. *Land Use Policy, 30*(1), 114–119. https://doi.org/10.1016/j.landusepol.2012.03.009.

Blomley, N. (2014, December). Making space for property. *Annals of the Association of American Geographers, 104*, 1291–1306. http://doi.org/10.1080/00045608.2014.941738.

Blomley, N. (2016). The territory of property. *Progress in Human Geography, 40*(5), 593–609. https://doi.org/10.1177/0309132515596380.

Brown, K. M., Flemsæter, F., & Rønningen, K. (2019, January). More-than-human geographies of property: Moving towards spatial justice with response-ability. *Geoforum, 99*, 54–62. http://doi.org/10.1016/j.geoforum.2018.12.012.

Cadieux, K. V. (2011). Competing discourses of nature in exurbia. *GeoJournal, 76*, 341–363. https://doi.org/10.1007/s10708-009-9299-0.

Cooke, B., & Lane, R. (2015). How do amenity migrants learn to be environmental stewards of rural landscapes? *Landscape and Urban Planning, 134*, 43–52. https://doi.org/10.1016/j.landurbplan.2014.10.006.

Cooke, B., & Lane, R. (2018). Plant-human commoning: Navigating enclosure, neoliberal conservation, and plant mobility in exurban landscapes. *Annals of the*

American Association of Geographers, 108(6), 1715–1731. https://doi.org/10.1080/24694452.2018.1453776.

Crabtree, L. (2017). Transitioning around the elephant in the room: Resilience, property and the fate of cities. *City, 21*(6), 883–893. https://doi.org/10.1080/13604813.2017.1407583.

Fiege, M. (2005). The weedy west: Mobile nature, boundaries, and common space in the Montana landscape. *The Western Historical Quarterly, 35*(1), 22–47. Retrieved from http://www.foresthistory.org/Fellowships/Fiege.pdf%5Cnpapers2://publication/uuid/F675736D-D616-4290-BA9D-474D903ED11D.

Gammage, B. (2012). *The biggest estate on earth: How Aborigines made Australia*. Sydney: Allen & Unwin.

Graham, N. (2011). *Lawscapes: Property, environment and law*. London and New York: Routledge-Cavendish.

Hepburn, S. (2011). *Not quite The Castle: Why miners have a right to what's under your land*. The Conversation. Retrieved from https://theconversation.com/not-quite-the-castle-why-miners-have-a-right-to-whats-under-your-land-4176.

Hodgetts, T. (2018). Connectivity as a multiple: In, with and as "nature". *Area, 50*(1), 83–90. https://doi.org/10.1111/area.12353.

Jakes, A. F., Jones, P. F., Paige, L. C., Seidler, R. G., & Huijser, M. P. (2018, June). A fence runs through it: A call for greater attention to the influence of fences on wildlife and ecosystems. *Biological Conservation, 227*, 310–318. http://doi.org/10.1016/j.biocon.2018.09.026.

Keenan, Sarah. (2014). *Subversive property: Law and the production of spaces of belonging*. Abingdon, Oxon: Routledge.

Li, T. M. (2014). What is land? Assembling a resource for global investment. *Transactions of the Institute of British Geographers, 39*(4), 589–602. https://doi.org/10.1111/tran.12065.

Linebaugh, Peter. (2008). *The magna carta manifesto: Liberties and commons for all*. Oakland: University of California Press.

Mansfield, B. (2008). Property and the remaking of nature-society relations. In B. Mansfield (Ed.), *Privatization: Property and the remaking of nature-society relations* (pp. 1–13). Malden: Blackwell.

Marshall, G. R., Coleman, M. J., Sindel, B. M., Reeve, I. J., & Berney, P. J. (2016). Collective action in invasive species control, and prospects for community-based governance: The case of serrated tussock (Nassella trichotoma) in New South Wales, Australia. *Land Use Policy, 56*, 100–111.

Massey, D. (1980). The pattern of landownership and its implications for policy. *Built Environment, 6*(4), 263–271. Retrieved from http://www.jstor.org/stable/23284724.

Mendham, E., & Curtis, A. (2010). Taking over the reins: Trends and impacts of changes in rural property ownership. *Society & Natural Resources, 23*(7), 653–668. https://doi.org/10.1080/08941920801998893.
Ojalammi, S., & Blomley, N. (2015). Dancing with wolves: Making legal territory in a more-than-human world. *Geoforum, 62*, 51–60. https://doi.org/10.1016/j.geoforum.2015.03.022.
Pecl, G. T., Araújo, M. B., Bell J. D., Blanchard, J., Bonebrake, T. C., Chen, I. C., … Williams, S. E. (2017, in press). Biodiversity redistribution under climate change: Impacts on ecosystems and human well-being. *Science*. http://doi.org/10.1126/science.aai9214.
Yung, L., & Belsky, J. M. (2007). Private property rights and community goods: Negotiating landowner cooperation amid changing ownership on the Rocky Mountain Front. *Society & Natural Resources, 20*(8), 689–703. https://doi.org/10.1080/08941920701216586.

CHAPTER 3

Learning Conservation Practice in Rural-Amenity Landscapes

Abstract This chapter explores landholders' stories of conservation practice through learning-by-doing, as these stories offer a unique insight into the emergence of enviromental stewardship and the agency of non-humans in learning processes. Environmental stewardship also offers us a way to discuss how landholders' conservation practices can develop into certain styles or dispositions over time. This chapter unpacks two distinct dispositions that emerged through situated encounter; (1) an active or hands-on approach to conservation, and (2) a passive or hands-off approach to conservation. In interrogating conservation as a human-environment relationship, this chapter provides a novel perspective on the application of experiential learning in conservation more broadly.

Keywords Experiential learning · Stewardship · Conservation practice · More-than-human · Agency

1 Introduction

How private landholders learn to participate in conservation practice is critical for understanding the kinds of environmental stewardship that are emerging. This chapter explores landholders' stories of learning-by-doing, as these stories offer a unique insight into the co-constitution of conservation practice. We define stewardship broadly as the way that landholders

B. Cooke and R. Lane, *Making Ecologies on Private Land*,
https://doi.org/10.1007/978-3-030-31218-3_3

understand responsible care of the ecologies on their property in order to achieve both public and private benefits (Gill, 2013; Gill, Klepeis, & Chisholm, 2010; Worrell & Appleby, 2000). Environmental stewardship offers us a way to discuss how landholders' conservation practices can develop into certain styles or dispositions for conservation. To explore how human-environment interactions inform conservation practice, we interrogate the way that stewardship emerges through interactions and relations with nonhuman agents, particularly plants.

This chapter examines how rural-amenity migrants bring a history of human-environment relations with them when they in-migrate, and how the memory of those past interactions shape their contemporary stewardship. We also unpack the powerful role of people's early conservation practices on their property in shaping certain dispositions for stewardship that can then endure through time. As we discuss, these formative conservation practices are often connected to expressions of ownership of land. To help explain this process of learning to be environmental stewards we adapt a traditional experiential learning schematic (Kolb, 1984) to include nonhuman agency and demonstrate how this learning produces durable dispositions for environmental stewardship.

2 Environmental Stewardship and Experiential Learning with Nonhumans

While rural-amenity migrants can be highly motivated to pursue conservation practices, they may have limited familiarity with rural landscapes and be uncertain about the likely outcomes of their efforts (Mendham, Curtis, & Millar, 2012). As we noted in Chapter 1, amenity migrants can be unaware of local weed species (Klepeis, Gill, & Chisholm, 2009) and have minimal experience with practical land management tasks like fencing (Curry, Koczberski, & Selwood, 2001). Moreover, their aspirations to be environmental stewards can be bound up with other land use motives, like recreation (Urquhart & Courtney, 2011) or improving visual amenity (Knoot, Schulte, & Rickenbach, 2010). In this sense, rural-amenity migration is a complex socio-ecological transformation of the landscape, which produces significant uncertainty in terms of our understanding of how environmental stewardship emerges.

The diverse land use interests of rural-amenity migrants and their often limited hands-on conservation experience suggests that they will build on their initial ideas of environmental stewardship as they establish their amenity lifestyles. In farming contexts, stewardship has been shown to have a strong emergent dimension over time, linked to interaction with the physical landscape and knowledge passed between generations of farmers and amongst farming communities (Berkes & Turner, 2006; Trigger, Toussaint, & Mulcock, 2010). As a consequence, the stewardship of farmers shapes the trajectories of ecological change and is a central consideration in environmental policy design and implementation (Greiner & Gregg, 2011; Sorice & Donlan, 2015).

A more in-depth understanding of the conservation practices of amenity migrants can offer critical insights into how non-productive rural landholders learn to be environmental stewards (Gill et al., 2010). In keeping with the focus of this book, the conservation practices of amenity migrants present a useful starting point for interrogating the emergence of environmental stewardship. Indeed, we are specifically interested in how stewardship emerges *through* the practice of caring for, interacting with and observing the landscape in which one is situated. How do these relational practices of conservation shape an expression of environmental stewardship? And what does attention to the agency of nonhumans in the emergence of stewardship show us in terms of how conservation is being conducted?

Firstly, both extensive research on experiential learning in natural resource management (NRM) and farming scholarship have highlighted the prominent role of learning-by-doing on the part of landholders when it comes to informing conservation and environmental management (Armitage et al., 2009). As ecologies change through human intervention and natural processes, trial and error becomes pivotal for learning how to respond to changing landscapes (Berkes & Turner, 2006). Secondly, the experiential learning practices of landholders present a clear avenue for enfolding nonhuman agency into an articulation of environmental stewardship. Recognising the agency of nonhumans shows how conservation practice is active and ongoing, and constituted between people and ecologies over time. As this chapter signals the beginning of the empirical section of the book, Table 1 introduces the landholders who were interviewed, their land uses and the duration of their land tenure.

Experiential learning theory is described as 'the process whereby knowledge is created through the transformation of experience' (Kolb, 1984, p. 41). According to Kolb, this kind of learning occurs through a continuous cycle of action and reflection. The active experimentation (immersion in the world) and reflective observation (observing processes and reflecting on the outcomes) dimensions of experiential learning are highly relevant to

Table 1 Details of landholders who participated in this study, including names (pseudonyms), land use, conservation practice and duration of property tenure

Participant/s	*Land use*	*Conservation practices*	*Property tenure*
Jenny	Hobby farming[a] with some conservation	Landscape restoration through planting linear fence lines; weed control; fencing	15 years
Janet	Hobby farming/conservation	Restoration of large forest patch in gully; general weed control; fencing	12 years
Kim and Terry	Conservation only	Large scale pine tree removal, general weed control; restoration planting	8 years
Colin	Conservation only	General weed control; pest animal trapping	7 years
Elise	Hobby farm/small orchard/conservation	Creek line restoration planting; general weed control; fencing	20+ years
Jasmine	Hobby farming/conservation	Restoration planting around remnant patch; fencing; weed control; covenant	20+ years
Ahmed	Hobby farming/conservation	Linear restoration planting; weed control	9 years
Faye and Michael	Hobby farming/conservation	Large scale weed removal from creek gully; restoration planting; general weed control	11 years
Devon	Productive farming (sheep, cattle)/conservation	Spot planting in remnant patch; general weed control	15+ years
Natalie	Hobby farming/conservation	Fencing; spot planting; general weed control	6 years
Jason	Productive farming (cattle)/conservation	Restoration planting; fencing; general weed control; covenant	26 years
Joel	Productive farming (cattle)/home business/conservation	Fencing; large scale weed removal (blackberries); spot planting	Not disclosed
Adam	Small scale cattle grazing/home business/conservation	Fencing; linear restoration plantings; general weed control; covenant	4 years
Felix	Small scale cattle grazing/conservation	Fencing; linear restoration planting; general weed control; spot planting; covenant	20+ years

Participant/s	*Land use*	*Conservation practices*	*Property tenure*
Kevin	Small scale cattle grazing/conservation	Fencing; spot planting; general weed control; covenant	28 years
Ingrid	Productive dairy farm/conservation	Large scale restoration; weed control; fencing; spot planting	20+ years
Jacinta	Lifestyle landholding/conservation	Large scale restoration planting; general weed removal	7 years
Wendy	Productive farming (mixed use)/conservation	Large scale restoration planting; weed control; fencing	12 years
Jim and Beatrice	Home business/conservation	Spot planting, exclusion fencing, weed control, pest animal trapping	15 years
Steve	Hobby farming/conservation	Weed control, spot planting	9 years
Kelly	Conservation only	Weed control and spot planting	11 years
Liz	Productive farming and conservation	Weed control	20 years
Rob	Conservation	Weed control, large scale restoration planting and spot planting	20+ years
Trevor	Retired farmer on formerly productive farming/conservation	Weed control, linear restoration plantings, spot planting, retention of large old trees in paddocks	20+ years
Alex and Simone	Home business/hobby farming/conservation	Linear tree planting, spot planting, weed control	14 years
Emma	Conservation only	Weed control (hand pulling and herbicide)	26 years
Sally	Hobby farming and conservation	Weed control, spot planting, linear tree planting	8 years
Karen	Hobby farming and conservation	Weed control, spot planting, fencing remnant vegetation	13 years
Ken	Hobby farming/conservation	Weed control (herbicide and large scale burning) and spot planting	6 years
Maddy	Hobby farming/retired farmer/conservation	Weed control, spot planting, fencing remnant vegetation, linear tree planting	8 years
Alice and Sam	Conservation only	Weed control	14 years

(continued)

Table 1 (continued)

Participant/s	*Land use*	*Conservation practices*	*Property tenure*
Pauline and Allan	Productive farming/conservation	Weed control, linear tree planting, fencing of remnant vegetation	12 years
Dan	Retired farmer on formerly productive farm	Weed control, linear, patch and creek line large scale restoration	28 years
William	Conservation only	Weed control and spot planting	17 years
Lauren	Productive farming/conservation	Weed control and fencing of remnant vegetation	22 years
Hannah	Hobby farming and conservation	Spot planting, weed control	18 years
Nick	Share cropping and conservation	Large scale restoration from seed, spot planting, weed control	7 years
Jeff and Claire	Hobby farming (scaled back formerly productive farm) and conservation	Weed control, exclusion fencing, creek line restoration, spot planting	14 years
Tina	Hobby farming and conservation	Weed control, large scale linear creek line restoration	22 years

[a]Hobby farming is defined here as a practice where some farming is taking place (usually a small number of livestock), but it is largely for recreational or landscape aesthetic reasons. Little to no income is derived from hobby farming from a livelihood perspective, but produce is often consumed or shared

conservation practice, which explains its adoption in the field of environmental management more broadly (Fazey, Fazey, Salisbury, Lindenmayer, & Dovers, 2006; Kolb, Boyatzis, & Mainemelis, 2001; Leeuwis, 2004; Pannell et al., 2006). However, in line with more recent critiques, experiential learning should not be considered a closed, cyclical process of individual experience, reflection and cognition (Lankester, 2013; Leeuwis, 2004; Seaman, 2008). Rather, experiential learning needs room to acknowledge and engage with a wider scope of human-environment relations. This means de-centring the model of the individual learner divorced from a social world that has afflicted experiential learning (see, for example, Kayes, 2002; Lankester, 2013 for an NRM context). However, nonhuman agency remains largely un-conceptualised as part of experiential learning. We argue that the 'ongoing togetherness' (Cloke & Jones, 2001, p. 651) of people and nonhumans needs to be more prominent as a dimension of emergent stewardship (Fig. 1).

Figure 1 shows how experiential learning can be positioned as an influence that interacts with imported land use aspirations and sociocultural context in shaping the emergent stewardship of amenity migrants. Imported land use aspirations reflect the lifestyle motives that amenity migrants bring with them (noted above), while social and cultural context reflects how the role of social interaction in rural communities and other forms of information exchange can contribute to landholders' emergent stewardship (Larsen & Hutton, 2011; Pannell et al., 2006). Alongside these elements are the interactions of people with their landscape—the ongoing and dynamic learning process that comes with landholders 'acting' on and with nonhumans through practice (weeding, planting etc.).

3 The Emergence of Stewardship Through Rural-Amenity Lifestyle

In many of the stories told by landholders, the 'amenity' they sought through their in-migration to a rural landscape was closely tied to the private property parcel. Many participants who had migrated from the city spoke of wanting a rural property that was secluded, with greater privacy and separation from neighbours than what they had enjoyed in their previous suburban neighbourhood. Some of the participants who came from suburban or rural residential backgrounds expressed a desire to be the autonomous custodians of a patch of land—wanting to 'own a bit of bushland' (Steve) was something many had long coveted. The desire for

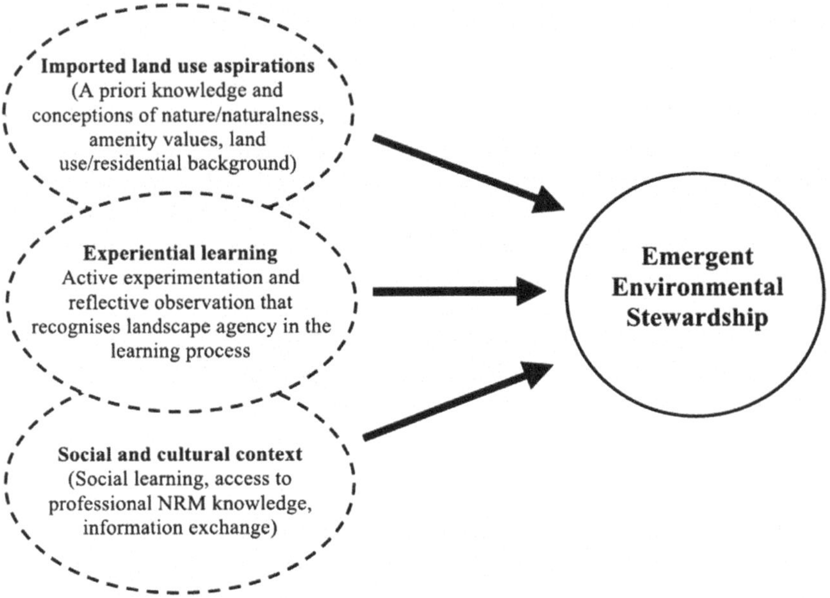

Fig. 1 Integrating nonhuman agency into learning about conservation practice, and the emergence of landholders' notions of stewardship. The dashed and overlapping lines reflect that the influences on emergent environment stewardship are not isolated and discrete, but open and inter-related

immersion in 'private nature' (Cadieux, 2011, p. 348) through property-based experiences is not unique to this cohort of rural-amenity migrants (Gill et al., 2010). The importance of the property as a space of sanctuary and nature engagement is a vital foundation for understanding why the property parcel is such a critical space for learning to be an environmental steward. Enthusiasm for getting their hands dirty from the outset meant landholders were tacitly engaged with plants from very early in their tenure.

The formative conservation practices conducted by landholders on their properties largely consisted of efforts to establish gardens and ornamental trees around the home. Concerns with gardens, landscape aesthetics and screening out neighbours through tree planting were indicative of their imported lifestyle aspirations. The planting itself was a kind of 'tinkering' (Jim) with the landscape, framed as an individual and recreational activity

(Cadieux, 2011; Urquhart & Courtney, 2011). The fact that some landholders had attempted to plant species that they had grown at their previous home, highlighted that past landscape associations and a priori knowledge were formative influences on conservation practices. Indeed, establishing a garden that connected to one's personal history served as an avenue for bringing a sense of familiarity to an unfamiliar landscape (Cadieux, 2011). William, Emma and Sally had all planted ornamental species (rhododendron, Japanese maple and silver birch, respectively) that featured in the gardens of a past suburban home. This reinforced the comfort provided by the ornamental garden; having a 'bush bit' and a 'home bit' (Emma) provided the 'best of both worlds' (William). This practice of establishing a familiar ecology appeared to be independent of the backgrounds of participants or the ecological characteristics of their property. Landholders like Jim and Beatrice, and Nick conducted similar plantings despite their different backgrounds and property ecologies (see table of landholder attributes). In seeking to create a 'homely' and familiar space (Power, 2009), many participants 'got planting' (Tina) almost immediately. Most of these early attempts at planting did not go as intended, with participants citing major failures:

> We tried to plant just a screening plantation [of mixed native and ornamental species] between our property and the next property. But what we've found is planting trees out here doesn't really work. I think the native trees seem to be strong enough and the others just don't survive... What is here is basically what has regenerated. (Hannah)

Hannah's struggle to establish ornamental trees reflected a formative experience amongst many participants—a realisation that their imported conservation knowledge and aspirations could not be easily applied to their new rural environment. The realisation that 'we really had no idea early on' (Steve) showed the emergence of an experiential learning that was embedded in experience of the property-centric ecology. And as Hannah outlined, this experience can quickly inform a conception of how (and how not) to be stewards of a particular piece of land. For many participants, it was the 'acting back' of ecologies in response to their intervention—especially through the agency of plants—that began to shape their particular knowledge of ecosystem function.

Hannah's attempts to plant non-native ornamental trees that ultimately died resulted in a view that only native species were 'strong enough' to survive in the landscape. Observing the natural regeneration of native species occurring in parallel with the death of the species she planted reinforced the idea that nature was best left to its own devices. The process of undertaking a conservation action and observing the response in the landscape developed specific ideas about ecological function and appropriate conservation practices. For William, the initial amenity motivation for living amongst 'native' nature was disrupted by the spread of non-native shrubs from an adjoining property. When attempting to remove these weeds by hand, William observed native birds nesting in them, causing him to question the benefits of removing all exotic species for local fauna. William had learnt through experience that you could be a counterproductive conservationist by aspiring to a 'pure' ecology of only native flora. Both William and Hannah typified how landholder practices could be a catalyst for generating ideas about ecological function and conservation: ideas that were grounded in property-based amenity.

The same experiential learning process was true for Natalie, who was a participant in the EcoTender conservation program that provided financial assistance for her conservation work. As Natalie explained:

> I did a massive replant in (my) first year. I put in, God knows... about 500 trees, but there would have been more because I planted for three years... I stopped and said 'This is really silly. There's something that's (regenerating) by itself, there's something that's coming up and I'm planting something in the middle'. So I basically stopped after that.

What is notable here is that Natalie had formed this perspective on how she should be stewarding her ecology prior to participating in EcoTender, which meant that she did not seek to do restoration planting as part of her involvement in the program. Rather, she limited her bid to covering the cost of fencing areas that could then be left to regenerate of their own accord. This is a clear example of how formative conservation practices can shape participation in conservation programs.

By generating ideas about ecological function through experiential learning, participants were developing an overarching disposition for environmental stewardship. We use the term 'disposition' here to refer to a durable mentality for stewardship that reflects a tendency to respond to one's environment in ways that reveal an ongoing association with that environment (Burton, 2012; Cammack, Convery, & Prince, 2011; Nordlund & Garvill, 2002). Through the kind of experiential learning processes already outlined, the environmental stewardship of participants developed as either a passive or active disposition. A passive disposition dictates a hands-off approach to stewardship, while an active disposition shapes a hands-on, interventionist approach. The active stewardship dispositions of landholders like Alice and Sam became evident through a discussion of the changes to vegetation communities on their property.

> *Alice*: So we've got all these bushes springing up. It's supposed to be native… but we're not sure.
> *Sam*: As they pop up we just pull them out… we just take them out because otherwise they'd take over.

When a protracted drought broke in south-eastern Australia in 2010, change began to occur to the bushland on Alice and Sam's property at a rate previously unseen. In the decade prior, Alice and Sam had come to appreciate a static bushland aesthetic on their property due to a long period of dry conditions that stunted plant growth and spread. However, pioneer acacia species (*Acacia pycnantha*) were suddenly flourishing, causing a noticeable disruption to this static aesthetic. In response, Alice and Sam removed the spreading acacia species to preserve the status quo ecology to which they had become accustomed (Fig. 2). Alice's admission that they were aware the species could be native (based on the observation of a similar plant in a local vegetation guidebook) also showed a prioritisation of experiential learning over other information sources—a point we return to later.

Active maintenance of a static ecosystem was a common practice amongst participants, with changing ecologies viewed as needing human intervention. As many landholders were only confident in identifying a handful of prominent weed species, the observed pioneering affordances of plants often served as a proxy for 'weediness'. The ways in which plants act as individuals but also collectively in their mobility across the landscape (Atchison & Head, 2013) played a powerful role in shaping conservation practices and thus, environmental stewardship. Plants acting in weedy ways—propagating quickly, spreading fast and 'taking over' (Steve)—were treated as a suspicious disruption to the ecology on their land.

Fig. 2 The natural regeneration of native acacia's (small shrubs in the foreground) resulted in efforts by Alice and Sam to remove them—a pile of recently cleared acacia's can be seen centre-right. A decade of living in a drought-affected landscape had generated a notion of the Australian bush as a static ecology

The emergence of passive stewardship dispositions was closely associated with the failure of trees planted by landholders to survive and thrive early in their property tenure.

> What I did try to do and it wasn't successful, plant some deciduous trees. I planted some oak trees down there… it'd be almost 20 years ago and they're no higher than about [one metre tall]. They just don't grow, so I gave up on that idea. So I don't plant any other trees and I haven't planted trees for years and years now. The [regenerating species are] managing all right. (Emma)

Emma attributed her struggle to grow introduced species to poor soil and inconsistent rainfall, something to which she perceived local species were well adapted. Thus, while Emma's ornamental trees remained stunted (Fig. 3), native bushland regenerated all around them. Just as Hannah had

Fig. 3 The contrast between the stunted growth of the Japanese maple (foreground) planted by Emma, compared to the flourishing regeneration of surrounding bushland, cemented Emma's passive stewardship disposition

experienced above, observing the contrast between the fortunes of native vegetation, compared to the trees Emma had planted herself, informed a view that native species 'belonged', and that she should remain a passive observer rather than an active participant in conservation practice. While landholders with passive stewardship dispositions undertook fewer conservation activities, they tended to be more comfortable with dynamic ecosystem change for the most part, which stood in contrast to the efforts to preserve static ecologies described earlier.

Two landholders in particular derived a unique kind of passive stewardship from their experiential learning through fear of not knowing 'the right thing to do' (Sally). One of these landholders, Sally, had a patch of trees die shortly after planting them. This process fuelled a belief that she should remain 'conscious not active' when it came to conservation practice, as she is likely to do more harm than good. Gareth, an NRM extension officer that worked in one of the case study areas, had encountered this perspective frequently in his dealings with rural-amenity migrants:

> I'll come across landholders who are passionate about "their bushland", but are sort of still in awe of it, and don't really interact with it. They'll walk the paths, but they'll talk about "oh no, we don't go in there because it's pristine, we don't want to touch that"… And I say, no, actually it's OK. You can go in to there, and indeed it's a really good thing, because you'll see what's going on. (Gareth)

As Gareth's experience indicates, the prospect of failure or of creating a disturbance as a deterrent to pursuing conservation practices may well be more widespread than landholder stories here indicate. And as has been raised elsewhere, there is a concern that these attitudes can deter rural-amenity migrants from pursuing vital practices like controlling weed spread. What is also notable from Gareth's account is the possessive attitude towards a property-centric ecology. The 'pristine' ecology is 'their bushland', a bushland that does not just represent the protection of nature from destructive land uses, but a bushland that is engaged with, understood and expressed through the lens of ownership. Being a steward of an ecology in this sense is not just caring for nonhumans, but caring for the nonhumans over which ownership rights and responsibilities extend.

3.1 Social Relations and the Emergence of Stewardship Dispositions

The majority of landholders discussed sources of learning that occurred outside their trial-and-error practices, with local community environment groups, plant nurseries, books and plant guides, and neighbours all contributing to their stewardship dispositions. Four landholders were particularly thankful for 'invaluable' (Maddy) advice on tree planting from nurseries in particular. Some participants like Maddy and Tina, actively sought out external advice after their initial plantings failed, seeking information about suitable species and planting arrangements for their properties. Ken, Steve, and Jeff and Claire explained a similar process of seeking out advice after struggling to remove weeds on their property through hand-pulling and spraying techniques. Claire had been 'dead-heading' a weed species (cutting the flower off before it goes to seed) as a way of trying to control its spread—a practice she had carried over from an existing interest in gardening. Having observed continual spread of the weed species, Claire felt they were 'going to have to seek advice' about alternative techniques. As Larsen and Hutton (2011) found, amenity migrants appear to seek out information based on specific conservation needs they identify that relate to their individual circumstances.

When landholders were exposed to external advice or information about conservation practice, this advice was often tested against their experiential learning. As seen from Sam and Alice's experience earlier, the decision to remove native species progressed despite identifying the species as native in a local native plant guide, because lived experience of surrounding ecological change was prioritised over external information that conflicted with that experience. However, there was some evidence of landholders changing their conservation practice through social encounters. For example, both Alice and Simone took advice from neighbours on the types of tree species they should consider planting after struggling to get other trees established, while Emma stopped removing a native shrub species she believed was a weed (after observing its rapid spread) after being told by a local farmer that the species in question was good bird habitat. The potential for stewardship dispositions grounded in experiential learning to shift as a result of wider social engagement suggests that landholder stewardship is 'durable yet changeable' (Kasper, 2009, p. 316) in the face of contradictory advice. Stewardship dispositions appear capable of shifting over time in response to 'trigger events', like timely interactions with other sources of knowledge when conservation practices are not going to plan (Sutherland et al., 2012). Yet, the power of experiential learning meant that stewardship that emerged through tangible experience of nonhuman agency could also be resistant to external knowledge.

The prioritisation of experiential learning over learning through social interactions may be indicative of differences between the stewardship of farmers and rural-amenity migrants. As noted earlier, farmers' collective interest in maintaining productive rural landscapes can produce strong social norms of acceptable conservation practice and facilitate knowledge sharing (Yung & Belsky, 2007). In the case of rural-amenity migration, where landholders often understand conservation as an individual, property-centric pursuit, experiential learning may be a particularly strong influence on the way landholders learn to be environmental stewards.

4 The Character and Meaning of Stewardship Dispositions

Attempting to determine which of the passive or active dispositions best reflects notions of 'best practice' conservation is a difficult and potentially fraught task, given the complex histories of settler-colonial landscape modification in Australia. Moreover, whether a passive or active stewardship approach is preferable is heavily contingent on the ecologies in question and the form and character of nonhuman agency in each location. Rather than recommending one approach as superior for generating ecological benefits, we would suggest that there is a need to reflect on how stewardship dispositions emerge and solidify in place and in relationship with nonhuman others.

Time and space were particularly critical in the emergence of passive and active stewardship dispositions. In terms of space, the property parcel was central to landholder learning about ecological function. As a result, when efforts to establish garden plantings based on imported knowledge failed, the experience triggered a highly transformative experiential learning process that was entirely bounded by the property space. The learning derived on the property could become resistant to information and advice that came from outside this space and often related to ecological function at larger scales than the property parcel (Knapp & Fernandez-Gimenez, 2009). The influence of space on learning may be particularly acute when people engage with an unfamiliar ecology or respond to an unfamiliar ecological change event like drought; phenomena that are likely to increase in frequency and severity under climate change.

Emergent stewardship practices and dispositions were also bound up with the change and continuity of surrounding ecologies (Cloke & Jones, 2001). Learning through practice means that knowledge is often being generated as ecologies materialise in the landscape (Head & Muir, 2006). For Alice and Sam it took a full decade before the spread of pioneering shrub species prompted them to actively intervene. For landholders like Hannah and Emma, the death of early garden plantings meant experiential learning occurred rapidly. Whether experiential learning is immediate or gradual, the learning that can emerge from specific actions is capable of shaping a stewardship disposition that is then applied to a range of conservation scenarios.

By recognising the agency of the living and inert entities that constitute rural landscapes, we show how these landscapes are more than just a setting in which the discrete, cognitive learning of individuals is applied in the form of conservation actions. The ecologies themselves are active in the process of learning and can play a substantive role in what ultimately constitutes landholders' environmental stewardship. In the case of many of the landholder stories featured here, the ecologies of influence were primarily those that persisted inside the bounds of their private property. Attention to nonhuman agency provides a much-needed avenue for expanding experiential learning beyond a depiction of the self-contained individual learner (Lankester, 2013; Seaman, 2008).

Our understanding of how rural-amenity migrants learn to be environmental stewards has implications for how they respond to and interact with conservation policy and programs. It is clear that early engagement with landholders is important for connecting their property-centric interests with landscape-scale conceptions of conservation, before their stewardship dispositions become too resistant to external advice. As has been noted elsewhere, landholders involved in environmental programs often like to understand how their participation in a program fits within the 'bigger picture' of conservation action at a landscape scale (Cocklin, Mautner, & Dibden, 2007). Outlining the bigger picture may enable more discussion about landscape-scale ecological function and collective conservation efforts with landholders, helping to counter issues like Sam and Alice's suspicion of dynamic ecological change. Mechanisms for sharing conservation practices and information amongst rural-amenity—potentially in the form of an ecological monitoring network—may be a practical way that an understanding of the bigger conservation picture can be formulated (see further discussion of this in Chapter 7). Early engagement with an extension officer also establishes a potential ongoing source of information about conservation for landholders, as they encounter various challenges during their property tenure.

The site of policy engagement with landholders is critical given that advice may be counter to the stewardship dispositions that have emerged through experiential learning. Engagement with landholders may be most effective in the property space, where those ideas originally emerged. As the walking method utilised in this research and described in Chapter 1 shows, the physical environment can be an important catalyst for conservation discussions as it embodies the experiences of landholders in that space over time. Engaging landholders in conservation discussions in the property space could help to encourage subtle shifts in durable stewardship dispositions, while also giving extension officers a better understanding of how and why they might possess certain perspectives on conservation practice. On-property encounters could also help overcome the hesitancy of landholders like Sally, whose passive stewardship resulted from a fear of doing the wrong thing. While passive stewardship can mean landholders are more accepting of dynamic ecological change, there is a risk that their hands-off approach will lead to benign neglect of local ecologies (Gill et al., 2010).

5 Conclusion

Our exploration of how stewardship dispositions emerge amongst amenity migrants offers an insight into the powerful interrelationship of nonhuman agency and experiential learning in shaping conservation practice. In interrogating this human-environment relationship, we offered a different perspective on the application of experiential learning in conservation more broadly. The stories of landholders in this chapter focused on the ways particular nonhumans (namely plants) participate in conservation practice. It should be noted though that this is but one form on nonhuman participation—there are many other nonhuman actors producing ecologies that people are often much less attuned to. Moreover, the lives of plants themselves are deeply entangled and dependent on other nonhuman actors, like soil and insects. In the following chapter, we delve further into the kinds of ecologies that are being produced through the conservation practices articulated here, and the ways that human-environment interactions are becoming embodied in landscapes over time. We are particularly interested in how the histories and legacies of human-environment interactions shape a trajectory for conservation practice in the present and future.

References

Armitage, D. R., Plummer, R., Berkes, F., Arthur, R. I., Charles, A. T., Davidson-Hunt, I. J., … Wollenberg, E. K. (2009). Adaptive co-management for social–ecological complexity. *Frontiers in Ecology and the Environment, 7*(2), 95–102. https://doi.org/10.1890/070089.

Atchison, J., & Head, L. (2013). Eradicating bodies in invasive plant management. *Environment and Planning D: Society and Space, 31*(6), 951–968. https://doi.org/10.1068/d17712.

Berkes, F., & Turner, N. J. (2006). Knowledge, learning and the evolution of conservation practice for social-ecological system resilience. *Human Ecology, 34*(4), 479–494. https://doi.org/10.1007/s10745-006-9008-2.

Burton, R. J. F. (2012). Understanding farmers' aesthetic preference for tidy agricultural landscapes: A Bourdieusian perspective. *Landscape Research, 37*(1), 51–71.

Cadieux, K. V. (2011). Competing discourses of nature in exurbia. *GeoJournal, 76*, 341–363. https://doi.org/10.1007/s10708-009-9299-0.

Cammack, P. J., Convery, I., & Prince, H. (2011). Gardens and birdwatching: Recreation, environmental management and human-nature interaction in an everyday location. *Area, 43*(3), 314–318. https://doi.org/10.1111/j.1475-4762.2011.00992.x.

Cloke, P., & Jones, O. (2001). Dwelling, place, and landscape: An orchard in Somerset. *Environment and Planning A, 33*(4), 649–666. https://doi.org/10.1068/a3383.

Cocklin, C., Mautner, N., & Dibden, J. (2007). Public policy, private landholders: Perspectives on policy mechanisms for sustainable land management. *Journal of Environmental Management, 85*, 986–998. https://doi.org/10.1016/j.jenvman.2006.11.009.

Curry, G. N., Koczberski, G., & Selwood, J. (2001). Cashing out, cashing in: Rural change on the south coast of Western Australia. *Australian Geographer, 32*(1), 109–124. https://doi.org/10.1080/00049180020036268.

Fazey, I., Fazey, J. A., Salisbury, J. G., Lindenmayer, D. B., & Dovers, S. (2006). The nature and role of experiential knowledge for environmental conservation. *Environmental Conservation, 33*(01), 1. https://doi.org/10.1017/S037689290600275X.

Gill, N. (2013). Making country good: Stewardship and environmental change in central Australian pastoral culture. *Transactions of the Institute of British Geographers, 39*(2), 265–277. https://doi.org/10.1111/tran.12025.

Gill, N., Klepeis, P., & Chisholm, L. (2010). Stewardship among lifestyle oriented rural landowners. *Journal of Environmental Planning and Management, 53*(3), 317–334. https://doi.org/10.1080/09640561003612890.

Greiner, R., & Gregg, D. (2011). Farmers' intrinsic motivations, barriers to the adoption of conservation practices and effectiveness of policy instruments:

Empirical evidence from northern Australia. *Land Use Policy, 28*(1), 257–265. https://doi.org/10.1016/j.landusepol.2010.06.006.
Head, L., & Muir, P. (2006). Suburban life and the boundaries of nature: Resilience and rupture in Australian backyard gardens. *Transactions of the Institute of British Geographers, 31,* 505–524.
Kasper, D. V. S. (2009). Ecological habitus: Toward a better understanding of socioecological relations. *Organization & Environment, 22*(3), 311–326. https://doi.org/10.1177/1086026609343098.
Kayes, D. C. (2002). Experiential learning and its critics: Preserving the role of experience in management learning and education. *Academy of Management Learning & Education, 1*(2), 137–149.
Klepeis, P., Gill, N., & Chisholm, L. (2009). Emerging amenity landscapes: Invasive weeds and land subdivision in rural Australia. *Land Use Policy, 26*(2), 380–392. https://doi.org/10.1016/j.landusepol.2008.04.006.
Knapp, C. N., & Fernandez-Gimenez, M. E. (2009). Knowledge in practice: Documenting rancher local knowledge in Northwest Colorado. *Rangeland Ecology & Management, 62*(6), 500–509. https://doi.org/10.2111/08-175.1.
Knoot, T. G., Schulte, L. A., & Rickenbach, M. (2010). Oak conservation and restoration on private forestlands: Negotiating a social-ecological landscape. *Environmental Management, 45*(1), 155–164. https://doi.org/10.1007/s00267-009-9404-7.
Kolb, D. A. (1984). *Experiential learning: Experience as the source of learning and development.* Englewood Cliffs: Prentice-Hall.
Kolb, D. A., Boyatzis, R. E., & Mainemelis, C. (2001). Experiential learning theory: Previous research and new directions. In R. J. Sternberg & L. F. Zhang (Eds.), *Perspectives on cognitive, learning, and thinking styles* (pp. 227–247). Mahwah, NJ: Lawrence Erlbaum.
Lankester, A. J. (2013). Conceptual and operational understanding of learning for sustainability: A case study of the beef industry in north-eastern Australia. *Journal of Environmental Management, 119,* 182–193. https://doi.org/10.1016/j.jenvman.2013.02.002.
Larsen, S., & Hutton, C. (2011). Community discourse and the emerging amenity landscapes of the rural American West. *GeoJournal, 77*(5), 651–665. https://doi.org/10.1007/s10708-011-9410-1.
Leeuwis, C. (2004). *Communication for rural innovation: Rethinking agricultural extension.* Oxford: Blackwell Science Ltd.
Mendham, E., Curtis, A., & Millar, J. (2012). The natural resource management implications of rural property turnover. *Ecology and Society, 17*(4), art5. https://doi.org/10.5751/es-05071-170405.
Nordlund, A. M., & Garvill, J. (2002). Value structures behind proenvironmental behavior. *Environment and Behavior, 34*(6), 740–756. https://doi.org/10.1177/001391602237244.

Pannell, D., Marshall, G. R., Curtis, A., Barr, N., Vanclay, F., & Wilkinson, R. (2006). Understanding and promoting adoption of conservation practices by rural landholders. *Australian Journal of Experimental Agriculture, 46*(11), 1407. https://doi.org/10.1071/EA05037.

Power, E. R. (2009). Border-processes and homemaking: Encounters with possums in suburban Australian homes. *Cultural Geographies, 16*(1), 29–54.

Seaman, J. (2008). End of the "learning cycles" era. *Journal of Experiential Education, 31*(1), 3–18.

Sorice, M. G., & Donlan, C. J. (2015). A human-centered framework for innovation in conservation incentive programs. *Ambio, 44*(8), 788–792.

Sutherland, L.-A., Burton, R., Ingram, J., Blackstock, K., Slee, B., & Gotts, N. (2012). Triggering change: Towards a conceptualisation of major change processes infarm decision-making. *Journal of Environmental Management, 104*, 142–151. https://doi.org/10.1016/j.jenvman.2012.03.013.

Trigger, D., Toussaint, Y., & Mulcock, J. (2010). Ecological restoration in Australia: Environmental discourses, landscape ideals, and the significance of human agency. *Society & Natural Resources, 23*(11), 1060–1074. https://doi.org/10.1080/08941920903232902.

Urquhart, J., & Courtney, P. (2011). Seeing the owner behind the trees: A typology of small-scale private woodland owners in England. *Forest Policy and Economics, 13*(7), 535–544. https://doi.org/10.1016/j.forpol.2011.05.010.

Worrell, R., & Appleby, M. C. (2000). Stewardship of natural resources: Definition, ethical and practical aspects. *Journal of Agricultural and Environmental Ethics, 12*, 263–277.

Yung, L., & Belsky, J. M. (2007). Private property rights and community goods: Negotiating landowner cooperation amid changing ownership on the Rocky Mountain front. *Society & Natural Resources, 20*(8), 689–703. https://doi.org/10.1080/08941920701216586.

CHAPTER 4

Landscape Legacy and the Making of Rural-Amenity Ecologies

Abstract In this chapter we focus on the temporality of emerging ecologies and conservation practices in rural-amenity landscapes. We explore how rural-amenity ecologies are being produced over time through interactions between people and plants. This focus includes some in-depth narratives of landholder conservation practice. We build on the insights from the previous chapter on how landholders learn about conservation practice to consider how human-environment interactions become embodied in the landscape. Our aim is to demonstrate the temporal trajectory and structuring influence of landscapes when it comes to current and future conservation practice. The way in which past human-environment interactions translate into the present and future will need to be carefully considered if we are to navigate a more reflexive approach to conservation practice.

Keywords Legacy · Ecologies · Landscape change · Conservation · Private land

1 Introduction

Rural-amenity migration is producing new ecologies, as regions that were once the domain of productive agriculture are now encompassing a suite of aesthetic, recreational and lifestyle land uses. To date, the implications

B. Cooke and R. Lane, *Making Ecologies on Private Land*,
https://doi.org/10.1007/978-3-030-31218-3_4

of these ecologies for conservation have been framed largely in 'traditional' conservation biology terms. This framing has focused on the threats or opportunities presented by rural-amenity migration for conservation or restoration of 'natural' environments to a past ecological benchmark—an imagined 'native nature'. We argue that the long history of human modification that characterises rural ecologies, combined with uncertainty about future ecological trajectories in light of global environmental change processes, necessitates rethinking the broader goals for conservation in rural-amenity landscapes.

In keeping with an emphasis on plants as nonhuman actors of interest, we explore how rural-amenity ecologies are being produced *over time* through interactions between people and plants. This focus builds on the insights from the previous chapter on how landholders learn about conservation practice to consider how human-environment interactions become embodied in the landscape. Our aim is to establish a temporal trajectory for current and future conservation practice. The way in which past human-environment interactions translate into the present and future will need to be carefully considered if we are to navigate a more reflexive approach to conservation practice.

A critical understanding of ecologies as *emergent* products of human and nonhuman interactions over time underpins this book (Castree, 2014). This interpretation is not only based on the provocation of the 'Anthropocene' era as a new epoch where humans are the driving force of planetary change (Castree, 2014; Robbins & Moore, 2013), but on the recognition that in settler-colonial contexts like Australia, there is no 'pre-human' ecology to which conservation can aspire (Head, 2011). These two realisations are extremely relevant for rural-amenity landscapes, given the often complex and multiple successions of landscape modification they have embodied over time, and the aspirations for conservation that their inhabitants possess.

As we established in Chapter 3, the intimate interactions between people and the environments in which they live are an essential starting point for interrogating rural-amenity ecologies. As Halfacree (2006, p. 309) notes, 'relationships between land and everyday life' are critical in the making and remaking of rural landscapes over time. We argue that the interrelationship between people, ecologies and *time* in the production of rural-amenity landscapes has been under-examined (Abrams, Gill, Gosnell, & Klepeis, 2012; Gill, Klepeis, & Chisholm, 2010). It is 'time' that we are particularly interested in here, especially in terms of the way that histories of land use and landscape modification reveal themselves in the course of contemporary conservation practices.

To help with our conceptualisation of time as embodied in the landscape, we connect with the work of anthropologist Tim Ingold, who has sought to re-embed human perception, knowledge and experience into the worlds that people inhabit (Ingold, 2000, 2011). A core tenant of Ingold's work is the notion that human perception of the environment is indivisible from our experiences of that environment (Ingold, 2011). Ingold suggests that movement along lines of becoming defines human existence, as we traverse through the world as wayfarers, with other human and nonhuman inhabitants (Ingold, 2007, 2011). Thus, 'inhabiting' the world is not being situated in a static place, but rather traversing along a 'way of life' (Ingold 2011, p. 4). In this conception, our lives extend along lines that represent our involvement in the world. Where lines of different inhabitants intersect, the 'life of each becomes bound up with the other' (Ingold, 2011, p. 148). These lines of movement carry with them a history and memory that is brought to bear on the entanglements that we have with others. How does this deepen our understanding of conservation practice? By imagining conservation practice as an entanglement of lines that represent human and nonhuman lives, conservation practice is not a series of discrete actions, but rather multiple interactions that mesh together different lived experience and histories.

2 The Temporality of People–Plant Relationships

The form and arrangement of heavily modified rural-amenity ecologies presented the perfect canvas for many landholders to further their aspirations for 'bringing back' native ecologies to the landscape. Nearly half the participants stated explicitly that they thought their in-migration would allow the return of native ecologies to a rural landscape that had long been 'over-worked' through farming practices. Alan's declaration that, 'people spent decades clearing the land here and we'll spend decades revegetating it', captured the broad redemptive aspiration that many participants brought with them to their new surroundings. This number is perhaps not surprising when the majority of participants expressing this sentiment were interviewed in 2016 and were part of EcoTender, which often funded landscape restoration efforts. However, our exploration of their efforts to bring these imported aspirations to fruition revealed the extent to which landholders are not the only actors contributing to the making of new ecologies. Throughout this section, we heavily foreground the role of plant agency by showing the influence of plants in disrupting and re-orientating the ecological aspirations of human actors through time (Head & Atchison, 2008).

Despite being less than a decade old, the Sugar Gums (*Eucalyptus cladoclyx*) that are establishing in Trevor's paddock are beginning to express a characteristic form. The dense timber contained in their single straight trunks—abnormally straight for eucalypt species—are already supporting dense, deep-green foliage. Combined with the wide, outward growth habit of the limbs, the foliage helps to buffer the wind that blows across the open grassy plains. The relative flourishing of the Sugar Gums during this period of time is notable, in comparison to the fortunes of other species over the same period, as south-eastern Australia experienced a particularly harsh period of drought during the early 2000s. Endemic to regions of South Australia notable for a dry climate, Sugar Gums possess a tolerance to drought that is distinguished even amongst eucalypt species. Indeed, the contrasting fortunes of different eucalypt species during the drought, reflected by the death of at least one of the locally endemic mature Yellow Gums (*Eucalyptus leucoxylon*) in the foreground of Fig. 1, reveal how Sugar Gums came to occupy a section of Trevor's paddock.

The fenced section of paddock was originally constructed to exclude sheep, in the hope that Yellow Gums would regenerate from seed stores that were likely still present in the soil. In the 1990s, a local NRM group gave Trevor a small grant to facilitate this regeneration. As the endemic large tree species for the region, Yellow Gum restoration is encouraged. Conservation groups and NRM agencies in Victoria rarely advocate for Sugar Gum planting, given the species is not 'indigenous' to the region. Moreover, the dense timber yielded by Sugar Gums, which is prized for its fence posts and firewood, does not readily form good hollows for local wildlife (Hamilton, 2001). However, in the years following, virtually no Yellow Gums self-seeded, and the fenced area became choked with pasture grass that spread from adjoining paddocks. As the drought began to take hold, the lack of natural regeneration became a secondary concern, as the mature Yellow Gums died back. While the causes of the Yellow Gum dieback are likely to be multiple, it was the overlap with the drought period that proved significant in the remaking of this ecology.

Fig. 1 Sugar Gums (*Eucalyptus cladoclyx*), located rear centre and right of the image, are beginning to fill the open paddock space in the foreground, that had initially set aside for the natural regeneration of the local the Yellow Gum Species (*Eucalyptus leucoxylon*)

Dieback of the Yellow Gums was contrasted by the persistence of some large old Sugar Gums in Trevor's yard, which were planted by the previous property owner. These mature Sugar Gums remained healthy through the early years of the drought with little to no tending. Visible from the lounge room and casting shade across the pathway from home to shed, the imposing persistence of the Sugar Gums was an ingrained component of Trevor's everyday life. It was this persistence that saw Trevor plant the Sugar Gums that are visible in Fig. 1 (Chapter 3), in lieu of the lack of natural Yellow Gum regeneration. Trevor was aware of the attitude that conservation agencies had towards Sugar Gums, but his lived experience of the robust character of these trees through adverse conditions took precedence.

The persistence and durability of Sugar Gums in the landscape demonstrated the capacity of these plants to actively propagate to other parts of the property. Trevor's planting of Sugar Gums in a new part of the property produced a rural-amenity ecology that is novel in its arrangement and location, while also being structured by the continuity with previous trees of this species and with the actions of a past landholder. In this sense, the habitation of the previous landholder is being extended but also reinterpreted by Trevor in the making of a rural-amenity ecology. Through inhabiting the landscape with the Sugar Gums over time, Trevor's perspective on what types of species should be restored in the landscape shifted. The proliferation of Sugar Gums reinforces that trees often persist along a timeline that far exceeds that of human life, showing how people–plant relations can connect through multiple generations via the lifecycle of the plant (Lien & Davison, 2010).

3 Working with and Against Historical Trajectories of Land Use

Soil seed banks play an unseen role in the ecology of plant assemblages. The seeds of plant species can lay dormant in the soil for extended periods, germinating quickly in response to disturbances like fire, or changes in soil moisture (Wang, Jiang, Toshio, & Zhou, 2013). This applies to both native species and those that have been imported, such as the host of pasture grass varieties from Europe, Africa and the Americas. The soil seed bank of pasture grasses such as Cocksfoot, pictured in Fig. 2, can persist in the soil for up to four years, making it difficult to remove when attempting ecological restoration projects (a relatively short duration compared to many native seeds) (Future Farm Industries CRC, 2011). Over a century of grazing and cropping in the region pictured above suggests a soil seed bank dominated by pasture grass seed. In this sense, the history of landscape modification that is visible above the ground is reflected below the ground, in the soil profile.

While the history of landscape modification embodied in the soil is out of sight, it is not out of mind. When conducting ecological restoration projects like the one taking place on Nick's property, pasture grasses are sprayed with herbicide to knock-down the existing grasses and exhaust the seed bank. In this case, the initial knock-down proved successful, which paved the way for the sowing of native grass seeds and seedlings, along with the planting of canopy tree seedlings, to replicate the pre-colonial grassy forest vegetation community of the area. While heavy rain that followed the initial planting triggered growth of native grasses and trees, it also helped to

Fig. 2 A tussock of native Poa (*Poa labillardierei*) planted by Nick (lower centre image) as part of a restoration project struggles to establish itself amidst the surrounding pasture grasses, dominated by Cocksfoot (*Dactylis glomerata*) (long, pale seed head). This photo was taken around two years after the restoration project began

germinate the Cocksfoot from soil seed. This initial flush of growth from the seed bank resulted in competition with the native grasses for light, space and nutrients. While some Cocksfoot undoubtedly blew in on the wind from surrounding paddocks, the dense swathes of Cocksfoot that quickly established underscored its abundance in the soil seed store.

Despite the thick growth of Cocksfoot grasses, the Yellow Gums that had been sparsely planted across the site in line with the imagined grassy forest ecological restoration, established quickly and broke through the mat of grasses (Fig. 3), in contrast to the struggling native Poa's pictured in Fig. 3. The capacity of the Yellow Gums to overcome the choking constraints of the rejuvenating Cocksfoot proved a trigger event for the rural-amenity ecology on Nick's land. Experiencing this Yellow Gum growth led Nick to plant

Fig. 3 Nick planted additional Yellow Gums in a section of his restoration project as they were establishing successfully amongst the introduced pasture grasses. A mature eucalypt can be seen at the rear of the photo

more of these Gums as he felt they 'did well' and sometimes 'you've just got to plant the stuff that survives in the ground'. According to Nick, during an annual site visit, the conservation officer responsible for overseeing his restoration project voiced displeasure at the increased density of eucalypts, suggesting that the site no longer reflected a grassy forest ecology.

The rural-amenity ecologies emerging here reveal a tension between a soil seed bank that embodies a history of landscape modification, and the redemptive ecological restoration aspirations that Nick initially sought to realise. The seed store of Cocksfoot—a hidden and dormant form of plant agency—proved an immediate and powerful actor. The constant replenishment of the seed store over the decades of its cultivation, activated in a flush of growth, had shifted the trajectory of conservation in a period of less than two years. Just as notably, as Cocksfoot materialised, Nick's practices shifted from conservation aimed at bringing back a past ecology, to prioritising the species that 'did well' in this landscape. The prominence of the Yellow Gums rising above the grasses as we walked Nick's property gave a sense of how Nick's conservation practices would be affected by experiencing this growth over time; the capacity of the gums to persist and then flourish represented an avenue for a future ecological trajectory that worked with, rather than against, the historical ecology that was already given.

While Nick's weed seed was in the soil, other landholders encountered weeds that would blow into their property on the wind, or flow through via creeks or surface water during heavy rain. The source of this weed spread was not only from another space or place, but also another time. Some of the contemporary weed management issues being encountered were the result of land uses and primary production industries long since abandoned. Monterey pine (*Pinus radiata*) is a common plantation species in Australia, proving very popular for timber, pulp and paper products. It's propensity for self-seeding and spreading through bushland has led to its classification as a noxious weed, and to targeted eradication programs across multiple levels of government in Australia (Lien & Davison, 2010). As suburbs have grown around Melbourne and nearby regional centres, some former pine plantations have been subdivided into house lots for sale to rural-amenity landholders, given the increased value of the land for housing. After newly created house blocks are cleared prior to sale, pine seeds persist in the soil due to the many years of plantation cycles. Soil disturbances like clearing, house building and landscaping can be the ideal trigger for catalysing the germination of this seed stock. The effectiveness of conservation objectives and practices requires careful reflection on the suitability of plants considered for restoration at a given site, allowing for plant agency that may lie dormant in the soil.

While site preparation, like weed spraying, is common practice amongst conservation practitioners, the goal of making space for the recreation of a pre-colonial ecological assemblage is unrealistic and inappropriate. The suitability of restoration species and the assemblages in which they are planted will be partly dictated by the embodied histories of landscape change. Moreover, the challenges and unpredictability of conservation practice in these landscapes must be clearly communicated to rural-amenity migrants with conservation aspirations; the importance of communicating the uncertainty of future outcomes is no better exemplified by the success of Yellow Gums on Nick's property, and their dieback on Trevor's property.

Embracing uncertainty when dealing with landholders may increase their preparedness for the types of 'surprise' outcomes experienced by Nick.

While the arrangement of ecologies differed across the three cases described above, they generated unique actions and interventions by landholders, which exposed them to past people–plant relations. The long 'temporal arc' (Lien & Davison, 2010, p. 250) of the old Sugar Gums across generations helped to facilitate their replanting in a new arrangement. While the soil seed bank of Cocksfoot lay dormant for decades, it was its rapid and uniform materialisation that catalysed Nick's planting of eucalypts. These cases serve to highlight a challenge for conservation practice in the present moment; having identified the emergent dimension of ecologies, how should conservation practice respond practically and conceptually to the ways in which embodied histories of landscape modification are structuring future ecological arrangements?

4 Landscape Legacy—A Temporal Trajectory of Conservation Practice

The concept of 'landscape legacy' holds promise as a heuristic for conservation practice in highly modified rural-amenity landscapes, where attention to the influence of time is especially pertinent. Articulations of 'legacy' in general vernacular capture the idea of something being handed down from the past, with the potential to affect the future. As we have seen, rural-amenity migrants often want to leave their properties in better ecological condition than when they purchased them. Yet, efforts to realise this legacy involve conservation practices in heavily modified landscapes that embody a rich history of interactions—a meshing of the lines of life, as Ingold would have it. We can think about the way that landscape histories structure conservation practices as a process of inheriting a landscape legacy from the past. Thus, conservation practice represents a moment where the bequeathing and inheriting of landscape legacies become entwined, as future aspirations and past practices are united through the production of novel and often unanticipated ecologies (Fig. 4).

Ingold's metaphor of wayfaring, discussed earlier in the chapter, is instructive for considering how rural-amenity migrants are encountering the structuring influence of past landscape legacies. Following the landholder narratives above, we can more easily see how past people–plant relations become embodied in the landscape, 'lay(ing) a trail' (Ingold, 2011, p. 148) that becomes enmeshed with contemporary landholders' lives. This

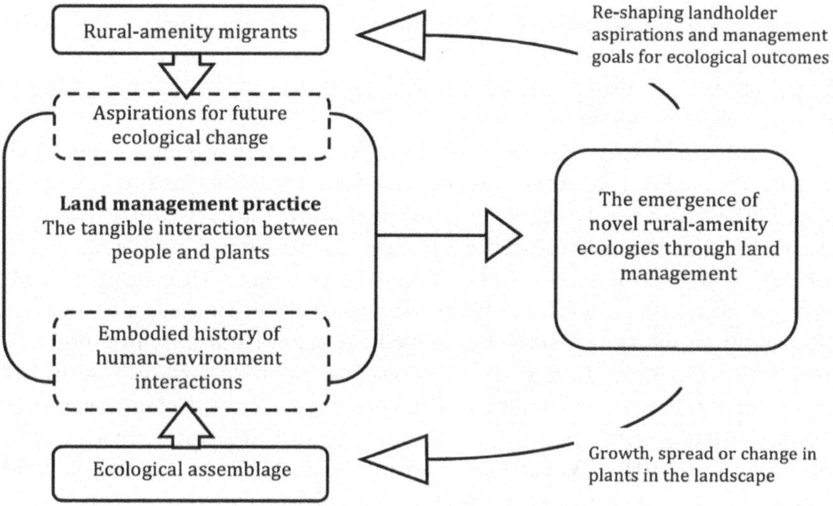

Fig. 4 Landscape legacy characterises the transformative process of conservation practice as a bringing together of past and future, as people and plants cohabit their surrounds and navigate the meshwork of embodied landscape histories in setting trajectories for future ecologies

meshwork of trails, or trajectories, structures the way that ecologies are produced through conservation practice. In this way, current landholders are performing within the pathways that have been set down by previous inhabitants. As we have seen, however, the trajectories of these trails are diffuse, and new paths can be forged as people and plants inhabit landscapes in ways that resist embodied legacies. Landscape legacy is not about a singular past being 'transmitted ready-made' (Ingold, 2011, p. 141) into the present, but trails of human and nonhuman interactions that set a course for current and future ecologies.

Relating landscape legacy back to the experiences of participants above, Trevor can be seen to have inherited a legacy through the Sugar Gums that were planted by the previous owner of his property. Over time, Trevor's inherited legacy was translated into a new legacy that is being projected forward through the growth of Sugar Gums in a new assemblage—a new trail, but one that has diverged from a past trajectory. The legacy of farming on Nick's property dictated an ecological trajectory that Nick initially resisted, before seeking to encourage a novel ecology that could persist within parameters set by past activity. Consequently, as amenity migrants traced the trajectories of past landscape interactions, they became co-participants in the production of future ecologies alongside latent seed banks and drought-resistant eucalypt species.

Evident in the experiences of all three landholders was that even when redemptive ecological conservation was their initial aspiration, the process of inhabiting rural ecologies—but most importantly, being affected by them—produced unanticipated ecological assemblages. While Trevor and Nick eventually embraced novel ecological assemblages through trial-and-error, we argue that effective conservation must allow for ecological novelty as a potential outcome. However, as we have seen, rural-amenity migrants are part of the unfolding story of ecological novelty through landscape legacy; *over time*, the enfolding of the lives of people and plants produces ecologies that are unique in their assemblage and function (Head, Atchison, & Phillips, 2014). We argue for an approach that positions private land conservation as an ongoing experiment involving multiple agents working together under uncertainty (Hinchliffe, 2008; Lorimer & Driessen, 2014; West, Beilin, Wagenaar, & Watkins, 2019). Thinking about the broader goals of private land conservation in experimental terms allows us to consider how best to create or maintain ecological form, process and function in ways that respond to ecological trajectories rather than ecological benchmarks. Preparedness amongst private land conservation managers and policy-makers to progress such experiments collaboratively, rather than prescribing goals of pre-colonial native ecologies, may produce future ecologies that are more capable of responding to larger scale ecological processes.

In acknowledging landscape legacy, we argue for more reflective private land conservation that considers whose legacy is present in the landscape, and what responsibilities contemporary landholders have to the continuity of those legacies. We specifically refer to the idea that in settler-colonial societies, Indigenous Peoples have managed, cared for and cultivated the land for many thousands of years (Pascoe, 2014). A conscious reflection on legacies necessitates listening to other ways of knowing and caring for the land as these ways are already enmeshed in the world in which conservation practices are taking place; we expand further on this in the conclusion chapter. Moreover, legacy might offer a heuristic for the kinds of responsibilities and obligations that need to be taken up as part of property ownership. In this sense, legacy could be used to link property rights with obligations to address First Nations' sovereignty and repatriations in settler-colonial contexts. Legacy can also serve a more general role in promoting notions of obligation that come with property ownership, so that the rights to use or exploit land in the present are countered by the responsibility to foster legacies that are embodied in the landscape through conservation practice.

5 Conclusion

The persistence of trees across generations of landholders and the dormancy and rapid propagation of soil seed banks have shown how the lives of plants encapsulate an extensive history of land use change. These altered landscapes structure a trajectory for ecological assemblages, as plants and plant interactions mediate the redemptive ecological aspirations of landholders. How to make space for the co-production of rural-amenity ecologies through conservation practice presents as a significant challenge.

Landscape legacy offers a way to conceive of how human and non-human actors are making and remaking rural-amenity ecologies. Rather than engaging with fixed historical baselines for conservation, landscape legacy positions conservation as a *process*, with people positioned within a temporal trajectory of ecological change. While these trajectories can be re-shaped and re-orientated, the embodiment of trajectories through the 'meshwork' of people–plant relations over time serves as a pathway for the emergence of future ecological assemblages. Thus, landscape legacy allows us to see contemporary conservation practice as a confluence of inheriting and bequeathing legacies through the co-habitation of people and plants. Such a perspective invites reflections for what open and reflexive conservation might look like into the future, and who might participate in legacy-making projects. Moreover, as raised in the previous chapter, there is a need to share practices and information about ecologies across private land, so that ways to enhance ecological diversity and abundance can be better understood collectively in rural-amenity landscapes. We draw out the needs and priorities for sharing of practices and monitoring processes in our final chapter (Chapter 7).

Because it demands a sensitivity to time, the idea of legacies as part of conservation practice served to establish a more sophisticated understanding of how people's conservation practices interact with government policies and programs. Acknowledging that people and landscapes become entangled through conservation practice allows the positioning of government-initiated conservation programs like covenants and payment schemes as another entity in that entanglement. This matters because it helps to move away from the idea that conservation programs are smoothly and uniformly enacted, and that their programmatic intent is realised in clear and unencumbered terms. Through landscape legacy, we also have a vehicle by which to articulate the obligations that accompany the rights of property ownership, and thus assist in efforts to re-imagine the relations of property discussed in Chapter 2. In the following chapter we expand on these ideas by exploring the ways that landholders enact conservation covenants on their properties, and how this enacting is tied up in the legacy of past land users and the potential future legacies for their own conservation practice.

References

Abrams, J., Gill, N., Gosnell, H., & Klepeis, P. (2012). Re-creating the rural, reconstructing nature: An international literature review of the environmental implications of amenity migration. *Conservation and Society, 10,* 270. https://doi.org/10.4103/0972-4923.101837.

Castree, N. (2014). The Anthropocene and geography I: The back story. *Geography Compass, 8,* 436–449. https://doi.org/10.1111/gec3.12141.

Future Farm Industries CRC. (2011). *Management guide to minimise environmental weed risk: Cocksfoot.* Retrieved from www.futurefarmonline.com.au/about/weedrisk.

Gill, N., Klepeis, P., & Chisholm, L. (2010). Stewardship among lifestyle oriented rural landowners. *Journal of Environmental Planning and Management, 53,* 317–334. https://doi.org/10.1080/09640561003612890.

Halfacree, K. (2006). From dropping out to leading on? British counter-cultural back-to-the-land in a changing rurality. *Progress in Human Geography, 30*(3), 309–336. https://doi.org/10.1191/0309132506ph609oa.

Hamilton, L. (2001). *The Sugar Gum story: The marketing success of a humble shelter tree.* The Regional Institute Ltd.

Head, L. (2011). Decentring 1788: Beyond biotic nativeness. *Geographical Research, 5,* 166–178. https://doi.org/10.1111/j.1745-5871.2011.00746.x.

Head, L., & Atchison, J. (2008). Cultural ecology: Emerging human-plant geographies. *Progress in Human Geography, 33,* 236–245. https://doi.org/10.1177/0309132508094075.

Head, L., Atchison, J., & Phillips, C. (2014). The distinctive capacities of plants: Re-thinking difference via invasive species. *Transactions of the Institute of British Geographers, 40*(3), 399–413.

Hinchliffe, S. (2008). Reconstituting nature conservation: Towards a careful political ecology. *Geoforum, 39,* 88–97. https://doi.org/10.1016/j.geoforum.2006.09.007.

Ingold, T. (2000). *The perception of the environment: Essays on livelihood, dwelling and skill.* London: Routledge.

Ingold, T. (2007). *Lines: A brief history.* Oxon, UK: Routledge.

Ingold, T. (2011). *Being alive: Essays on movement, knowledge and description.* London: Routledge.

Lien, M. E., & Davison, A. (2010). Roots, rupture and remembrance: The Tasmanian lives of the Monterey pine. *Journal of Material Cultures, 15,* 233–253. https://doi.org/10.1177/1359183510364078.

Lorimer, J., & Driessen, C. (2014). Wild experiments at the Oostvaardersplassen: Rethinking environmentalism in the Anthropocene. *Transactions of the Institute of British Geography, 39,* 169–181. https://doi.org/10.1111/tran.12030.

Pascoe, B. (2014). *Dark Emu: Black seeds: Agriculture or accident.* Broome: Magabala Books Aboriginal Corporation.

Robbins, P., & Moore, S. (2013). Ecological anxiety disorder: Diagnosing the politics of the Anthropocene. *Cultural Geography, 20,* 3–19. https://doi.org/10.1177/1474474012469887.

Wang, Y., Jiang, D., Toshio, O., & Zhou, Q. (2013). Recent advances in soil seed bank research. *Contemporary Problems of Ecology, 6,* 520–524. https://doi.org/10.1134/S1995425513050181.

West, S., Beilin, R., Wagenaar, H., & Watkins, C. (2019). Introducing a practice perspective on monitoring for adaptive management.*People and Nature, 1*(3), 387–405.

CHAPTER 5

Permanent Protection and the Legacies and Spatialities of Conservation Practice

Abstract Permanent protection through conservation covenants has emerged as a powerful mechanism for securing ecologies on private land. In this chapter, we explore how ideas of permanent protection are woven into the everyday practice of conservation examined in previous chapters. We explain how rural-amenity landholders perceive the work that their covenant is doing, and to unpack their uncertainties about the security of permanent protection, especially in light of the effect of surrounding land uses on their property. In exploring these themes, we examine differences between protecting ecological legacies through legally binding covenants, and the need to recognise and build on the legacy of experience, learning, labour and care for ecologies, as part of efforts to enable conservation outcomes beyond the tenure of a single landholder.

Keywords Covenants · Easements · Conservation · Nature · Permanent protection

1 Introduction

Permanent protection through conservation covenants has emerged as a powerful mechanism for securing ecologies on private land. Australia now has over 7500 conservation covenants (the vast majority on private land) which contribute around five percent of the area protected under the

B. Cooke and R. Lane, *Making Ecologies on Private Land*,
https://doi.org/10.1007/978-3-030-31218-3_5

National Reserve System (Hardy, Fitzsimons, Bekessy, & Gordon, 2017). Our aim in this chapter is to contribute to a small body of qualitative research on why landholders are interested in conservation covenants (Rissman & Sayre, 2012). But more than just understanding the appeal of permanent protection, we seek to explore how ideas of permanent protection are woven into the kinds of everyday practice of conservation that we explored in previous chapters. We use our relational understanding of conservation practice to explain how landholders perceive the work that their covenant is doing, and to unpack their uncertainties and anxieties about the limitations of permanent protection as an overarching ambition.

Rural-amenity landscapes offer a unique space in which to examine conservation covenants because landholders in these areas are often motivated to protect scattered ecological remnants they have invested significant effort in rehabilitating. The marginal and precarious existence of these rural-amenity ecologies brings an immediacy to the need for protection, but also highlights the challenges of maintaining the ecologies that have been restored. In exploring these themes, we examine differences between protecting ecological legacies through legally binding property covenants, and the need to recognise and build on the legacy of experience, learning, labour and care for ecologies that extend well beyond the boundaries of an individual private property.

In our introduction to covenants below, we outline the way they function in Australia, their growing role in contributing to international biodiversity conservation and how they are understood as a particular governance tool. The covenants examined here are those offered by Trust for Nature in Victoria, and those available to landholders through the Victorian *Conservation, Forests and Lands Act* (1994). We position covenants as part of a broader move to more neoliberal modes of environmental governance. We explain what we mean by neoliberal environmental governance below, and the role played by covenants in that process.

Our analysis hinges on the idea that the intentions of policy in the conservation field can sometimes be re-imagined and co-opted when it comes to actual, on-ground implementation. Indeed, there is a growing body of work that explores how policy intentions seldom map onto the world in neat and complete ways (Van Hecken, Bastiaensen, & Huybrechs, 2015). What is less clear is the means by which policies and programs might swerve from the desired path when they encounter the people and ecologies that are the intended subjects of these policies. This chapter contributes to this conversation through a detailed case study that explores how conservation interventions like covenants 'succeed or fail when they move from ideas into practice' (Stuart, Benveniste, & Harris 2014, p. 35). This kind of fine-grained analysis directs attention to the messy ways in which policy is enacted, and the agency of people and ecologies in shaping on-ground outcomes.

2 Covenants, Permanent Protection and Neoliberal Governance

In the context of private land, conservation covenants are legally binding agreements struck between landholders and governments or land trusts. These agreements revolve around the restriction or control of particular land uses such as grazing or property subdivision, in order to preserve some ecological asset (Hardy et al., 2017). In relation to the discussion of property as a 'bundle of rights' in Chapter 2, covenanters are effectively restricting some of their use and development rights voluntarily. Along with these restrictions, some covenants can also mandate specific management actions, such as the control of invasive flora or fauna. Most covenants are linked to the title or deed of the property in question and so run with the land not the property owner (Kay, 2016). This means they are designed to protect both current and future ecological values. Thus, the landholder who signs the agreement and all future landholders are required to uphold the stated obligations of the covenant.

In order to qualify for a Trust for Nature covenant, a property must have ecological values considered to be worthy of protection by the Trust. As with many land trusts, Trust for Nature prioritises ecologies that might be rare, threatened or under-represented in the National Reserve System. However, covenants signed under the *Conservation, Forests and Lands* Act (1987) that are featured in this chapter have a lower standard in terms of the 'intact-ness' of their ecologies. For example, landholders with such covenants may protect patches of restoration plantings that contain no remnant vegetation and are considered to be a common vegetation community in Victoria.

While conservation covenants have been around for many decades in Australia and overseas, they are becoming an increasingly popular conservation mechanism amongst both landholders and policy-makers (Bingham et al., 2017). In Victoria, Trust for Nature signed their one thousandth covenant in 2011 and over 1400 covenants were signed up to 2019. The numbers of covenants signed per year grew steadily from 1987, then escalated rapidly during the 2000s. This growth is not dissimilar to the experience elsewhere, with conservation easements now the most common mechanism for private land conservation in the United States (Morris, 2008). The International Union for Conservation of Nature (IUCN) recognises covenants on private land as a legitimate form of protected area, which allows governments to count private land under covenant towards their international obligations for biodiversity protection.

Increasing demand for rural property by rural-amenity landholders (or 'lifestyle' oriented landholders in more traditionally low amenity, productive landscapes) has been cited as one of the reasons for the increasing interest in covenanting in Australia (Fitzsimons & Carr, 2014). Less reliance on the property for income, and greater value ascribed to landscape aesthetics, means covenant land use restrictions are less confronting for those not relying on primary production as their main source of income (Moon & Cocklin, 2011). Clusters of covenants exist in rural-amenity landscapes to the east and south of Melbourne, as well as those around rural centres like Ballarat and Castlemaine to the north-west. This pattern of covenant uptake shows they are proliferating in areas associated with rural amenity lifestyle migration. Amenity landholders have also been reported to be pursuing covenants on their land for 'privacy and private amenity use… [and are] keen to protect it from future development' (Urquhart & Courtney, 2011, p. 539). This suggests that covenants in amenity areas may be used to support more than just ecological ambitions, including things like maintaining agricultural practices that are threatened by urbanisation.

The proliferation of covenants offered by government and land trusts alike exemplifies the devolution of environmental governance. In the west, this has manifested as a move away from direct state intervention like land purchase for conservation, in favour of individualised, and voluntary efforts that encompass private land (Morris, 2008). As Kay (2016, p. 510) notes, this is not just a devolution of responsibility, but a 'devolved approach to regulating and managing space', as covenants produce parcels of property-scale zoning and protection. In the United States and the UK, as well as Australia, recent decades have seen the retreat of the state and an increase in 'partnerships between state, private sector and non-governmental organisations' (Adams, Hodge, & Sandbrook, 2014, p. 581). Land trusts and covenanting have been a big part of this move. Indeed, as Kay (2016, p. 509) notes:

> Several accounts of the history of the land trust movement have recognized the direct relationship between the handicapping of federal land acquisition efforts and the growing popularity of approaches that leverage private property rights toward public goals.

While the growth of covenants has accorded with the emergence in the 1980s of a neoliberal governance model (described below), it is simplistic to suggest covenants are inferior to state-led conservation. For example, some landholders have sought to use covenants to address what they view as a failure of government to properly legislate for environmental protection (Morris, 2008). And much of the funding that supports covenant roll-out has traditionally come from government, let alone the fact that land trusts like Trust for Nature are statutory bodies whose existence depends on government legislation. As such, covenants are more accurately understood as a type of hybrid governance arrangement (Lockwood & Davidson, 2010) rather than wholesale privatised conservation. What they do show is that decision-making, planning and coordination for protected area conservation is being transferred from the state to non-government organisations and individual landholders (Kay, 2016; Morris, 2008).

The rise of covenants is thus inevitably linked with the rise of neoliberal governance. Neoliberal governance has traditionally been defined by increased marketisation, privatisation and deregulation, in which state intervention is reduced (Mansfield, 2009). However, more recent incarnations have seen state agencies take an active role in regulating through market processes, with the aim of increasing cost efficiency while pursuing government policy agendas. Unsurprisingly, this complex process of both deregulation and re-regulation has also characterised recent approaches to environmental governance (Lockie & Higgins, 2007; Stuart et al., 2014). As covenants show, the neoliberalisation of environmental governance in particular is characterised by the increase in privately owned conservation lands, the promotion of market-based programs and activities and the commodification and marketisation of the natural environment (Fletcher & Breitling, 2012; Fletcher & Büscher, 2017). In what follows, we will specifically focus on the ways private property is put to work in order to 'save nature', and how this governance approach is tangled up in landholder conservation practice.

The policy and political context we have outlined above cannot be separated from what is happening on the ground and as part of everyday conservation practice. Setting the broader political context helps us to locate landholder experiences and discuss a nuanced picture of conservation covenants. With this governance context in mind, we now look at how covenants are playing out on the ground.

3 Landscape-Scale Conservation Through Property-Based Action?

Given the prominence of stewardship in shaping conservation practice, it was not surprising that the idea of conservation covenants resonated with a number of the landholders. Eleven landholders already had a covenant, and four were considering one in the future, and all these individuals expressed a strong desire to steward their land and protect existing ecologies. Much of the drive for making their stewardship permanent was an awareness of how little vegetation and habitat existed in the local landscape. Indeed, Liz's desire to protect her patch was triggered by an aerial photo showing 'there wasn't much bush left locally', so protecting habitat on her property became all the more important. In this context, private property was seen to offer something of a refuge in which conservation and protection work could be undertaken without fear of intervention or disruption from outside. Yet, accompanying the conservation efforts that centred on their own patch, five of the landholders with covenants discussed their attempts to encourage others to adopt the practice of 'permanent protection'. These participants were of the view that if others in the region also signed up to a covenant, then through a process of agglomeration, the sum of their individual contributions would contribute to conservation at a larger scale than each individual property in isolation:

> Our immediate Eastern neighbours... actually bought a bush-block between our properties to add to their property... They're (in) the process... of covenanting that too... So that will make, with the Trust for Nature block further over... a pretty good start for a hub here. (Claire)

Creating a 'hub' of covenants that would protect a larger ecology was seen as a way of creating linkages through their property or from their property to nearby public land. Moreover, getting another local to sign up to a covenant constituted a long-term conservation gain in the form of permanent protections. From this perspective, advocating the benefits of permanent protection to neighbours with remnant bushland represented a form of shared conservation practice that (it was hoped) would have benefits beyond their individual property. This approach is very much in keeping with professional ecological advice on enabling ecological connectivity across the landscape. As Steve noted:

> In fact we managed to talk to our neighbour... and she's now put a Trust for Nature covenant on her [property] as well. We've tried to get our other neighbour to do the same. That would have taken the three [adjacent] bush blocks [under covenant].

Sharing the covenant idea appeared more common than actually sharing information about hands-on conservation practice itself. When it did occur, it included attempts to quell their neighbours' concerns about the limitations that covenants can place on land use or property value by demonstrating that their covenant had not changed the way they could use their property. Indeed, four landholders explicitly stated that they did not advocate specific conservation practices to their neighbours. As Janet noted, the autonomy that comes with land ownership is respected in her rural community. Janet described the right to do what you want on your land in terms of being a 'king in a grass castle'. Moreover, covenanters suggested they did not often work collaboratively on cross-boundary issues, as they were 'pretty busy just doing our own stuff' (Jim). Indeed, only Steve appeared active in sharing advice about weed management with neighbours, although Jeff and Janet had made occasional efforts to advise neighbours on controlling invasive fauna like foxes. This presents an interesting tension between individualised conservation practice at the property scale, and the desire to contribute to conservation benefits at a landscape scale. It also highlights how property rights can play a role in mediating the sharing of conservation practice.

Awareness of how a covenant could contribute to conservation beyond the scale of their property was important for individual landholder participation. Covenant holders had attempted to produce a collective conservation good "in bite sized chunks, where the material effects of individual action become visible and enduring' (Macnaghten, 2008, p. 81). The work that private property does here is to give landholders a sense of control over the delivery of those 'visible' and 'enduring' contributions. The flip side of course is that the sense of control and autonomy that allows covenanters to get on with conservation is the same sentiment that constrained them from seeking to influence the practice of their neighbours. These 'conservation in private' actions are a useful reminder of the unexpected ways that private property can help to materialise conservation practice for the delivery of public good outcomes. However, it also requires us to attend to how aspirations for these collective benefits are actually realised, and how they can be constrained or undermined.

Conservation schemes that facilitate individualised action through covenants may be particularly attractive to rural-amenity landholders as they facilitate a desire for 'getting on with it' on their property (Klepeis, Gill, & Chisholm, 2009, p. 323). As Jim noted, being busy with their own conservation practice—their 'own stuff'—had meant limited time for contributing to local, collective conservation groups. This may align with a perception that collective action initiatives require an unrealistic commitment of time and energy. It was clear from landholders like Steve and Claire that the flexibility of undertaking conservation work when and how you choose is tied up in the sense of fulfilment that comes from rural-amenity ownership.

4 Projecting Landscape Legacy into the Future

As discussed in the previous chapter, conservation practice is an inherently temporal pursuit, resulting in engagement with past landscape legacies and imagined futures. Chapter 4 revealed how mediated engagements with such legacies produce dynamic ecologies. Conservation covenants contribute to both past and future elements of legacy—they have a capacity to protect the past from destruction but also to engender future legacies that endure beyond the tenure of current landholders. The ability for covenants to provide protection in perpetuity is what enables this for landholders. Indeed, as J. K. Gibson-Graham, J. Cameron, and S. Healy (2013) highlight, collective conservation and community benefits can be achieved through property-centric covenants when they provide the conditions were shared benefits can unfold over time. In other words, even if the spatial footprint of a covenant's benefit are constrained by property, their temporal influence signifies a commitment to common good or public benefit outcomes.

Preserving flora and fauna in perpetuity was a core attraction for many landholders, as one might expect when signing a covenant. As Devon explained, 'it'll be something good to leave that block to posterity'. However, it was not just the ecological legacy that landholders wanted left to 'posterity', but what those ecologies embodied. As Devon had detailed through his story, getting his property ecology into 'good condition' in the first place had taken years of toil. Devon's ecologies represented an investment of time and energy expressed through conservation practices, which is what sustained the flourishing of those ecologies in the face of weed infestation, drought and the threat of bushfire. As Steve identified, he has not just placed a legal protection on his property, but his conservation practices have been orientated towards 'taking (the bush) back to... natural' through removing weeds and revegetating. A covenant was seen to present a mechanism for harnessing and securing the conservation practices of their work against any future unravelling should they sell the property (Selinske et al., 2016). As one of the program coordinators noted, contemplating property sale can be a trigger for landholders to sign up to a covenant, as they do not want their work to be undone by a subsequent owner:

> (I)t's the aspiration to see it protected once they're gone... landholders have generally looked after their property for a long period of time. And they are very precious about it. (Gareth)

External threats like urbanisation and property subdivision play a similar role in prompting landholders to sign up to covenants for fear their legacy could be threatened (Farmer, Knapp, Meretsky, Chancellor, & Fischer, 2011; Lai & Kreuter, 2011). While only one of the participants with a covenant had signed up with thoughts of selling in the immediate future, others found it reassuring to know that protection existed if they did sell. The eight landholders who indicated they might consider a covenant at a later date unanimously gave as their reason the desire to protect their own hard work. As Hannah put it, 'just so what's happened (on the property) is maintained'. For potential future covenanters like Hannah, there seemed no reason to pursue a covenant while they had no intention of selling, as they 'trusted' themselves to steward the property in a responsible manner. The protection offered by the covenant was only useful for limiting potential negative impacts by a future owner.

The power of this legacy preservation motive was reinforced by the stories of landholders who had pursued a *Conservation, Forests and Lands* Act covenant. These covenants had been offered as part of the EcoTender scheme we cover in Chapter 6. As part of the tender bidding process (explained in more detail in the next chapter), landholders were offered the opportunity to place a legal covenant over remnant vegetation or revegetation sites. They were encouraged to do so by the allocation of a higher assessment weighting to bids that included a covenant, meaning that such bids would have a greater chance of success. Of the 18 landholders involved in this study who were participating in EcoTender, seven had secured a covenant, but four of those landholders had used this mechanism in a manner that departed from the scheme's intention. Instead of using the covenant to enhance the competitiveness of their bid and increase the likelihood of receiving money for conservation work, these four landholders deliberately under-quoted the cost of works in their bids so as to increase the chance of getting a covenant. Participant 6 explained the process of underbidding to get a covenant:

> Oh, I just took a punt (on costing the bid), what I thought might get through, what it would cost, and what I was prepared to accept as a fair deal… I paid half of (the management costs).

The covenant, rather than the money to cover management costs, became the 'hook' that enticed these landholders to participate. Jasmine and Jason employed the same strategy of estimating the land management costs associated with their bid (e.g. tree planting or weed removal) and then only requested half of that money. The desire to protect these heavily modified remnants and recently planted revegetation further reinforced that landholders can seek to protect the legacy of their conservation labour as well as the conservation values on their land. Jasmine noted that the covenant will 'stop a developer coming in and knocking it all over' when they do eventually sell (Fig. 1). Or in the words of Jenny, 'I don't want somebody to come in in a future time and bugger it up again. That would be an incredible waste'. The sense that landholders could secure their conservation work against future threats gave them confidence that the ecological benefits they had helped to generate could be retained if they sold the property.

Three of the four participants who pursued this underbidding strategy had sought advice from Trust for Nature, who had informed them that the ecologies on their properties were not of sufficient value or size to qualify for protection. As such, EcoTender filled a gap in the permanent protection options previously available to them from either government or land trusts. Moreover, as Jason noted, the cost associated with the covenant process is substantial, so the opportunity to have your conservation legacy protected was an attractive proposition:

Fig. 1 A small strip of linear remnant vegetation that has been covenanted through EcoTender. Protecting this against a future landholder who may seek to extend paddock space on either side was a key motivation

> The covenants are the important thing. Because when I'm too old to do this and we sell this, the [advertisement] in the paper… will say, 'covenanted trees'. (Jason)

Sentiments like this from Jason showed how covenants were sometimes seen as a way of trying to project forward the control that landholders had over their property, and subsequently their control over those ecologies. Covenants represented an avenue for reducing the risk that 'all the work' (Beatrice) being done to steward the land would be undermined. Claire and Jeff were particularly thankful they had successfully navigated the three-year process of adopting their covenant with Trust for Nature:

> *Claire*: (W)e are probably moving towards leaving the place at some stage in the foreseeable (future).
> *Jeff*: That was another reason for the covenant too – to make it permanent.
> *Claire*: …(W)e're just so glad that we did the covenanting when we did. Really, really glad.

The importance of permanent protection was reinforced by Jeff and Claire's conservation work on the roadside near their property. Indeed, it was here that their conservation interest began. However, a process of degradation due to intensification of local agriculture had dramatically increased weed infestation on this roadside. The perceived unwillingness of the local council to address these management issues had resulted in a dim view of contributing to collective conservation efforts on public land:

> So yes, that is another reason why it just seemed easier to target in on our place and just stop fighting the fight out there that just doesn't… it was just getting too hard. (Claire)

This episode created a sense that time invested on projects over which they had limited control represented a risk in terms of the permanence of the ecological benefits. Especially given they had experienced the degradation of the roadside within the period of their own property tenure. Thus, working 'on our own place' (Claire), and protecting that work through a 'permanent' (Jeff) conservation covenant, presented a logical means of consolidating their legacy. Not only did the property come to be the centre of conservation action, it was around this interest that the engagement with covenants occurred. In other words, covenants offer a means for protecting 'my' propertied conservation legacy. But how permanent are these legacies? The fence line that denotes the property boundary does not isolate the property ecology from wider socio-ecological interactions. The idea that the covenants can almost act as a physical barrier that insulates a single property parcel from the vagaries and unpredictability of land use and ecological change beyond the fence is clearly flawed. Yet private property had come to be the centre of conservation action for many landholders and it was around this interest that engagements with covenants occurred.

For some landholders, the realisation that a covenant did not seal off their property from socio-ecological threats and disturbance came as somewhat of a shock. Indeed, the impact of this realisation compromised the notion that covenants offered control over their landscape legacy. While looking over the fence at the soil already washing down from her neighbour's recently planted turnips, Claire expressed a sense of 'helplessness' in recognition of the fat that the flows and mobility of nonhuman elements rarely respect private property boundaries or covenant agreements, challenging the illusion of control she had previously ascribed to conservation efforts focused on private property.

For all but one landholder with a covenant, land use threats from adjacent properties had led them to question just how 'permanent' the protections of a covenant actually were. Impacts like the movement of weeds across boundaries and the downstream impacts of erosion had triggered a realisation that their property could not be isolationed from the surrounding landscape. Perceptions of perpetual security were further ruptured by the realisation that their covenant did not necessarily facilitate the passing on the ecological knowledge they had developed over time to the next property owner. This demonstrates the inseparability of material landscapes from the dynamics underpinning their creation—the 'permanence' of the ecologies fostered by landholders is tied to their own presence in the landscape. This is especially true when the maintenance of these 'natural' landscapes requires so much active input. As such, a number of landholders realised that the experiential knowledge that they had generated would need to be passed on in some form if their landscape legacies were going to endure into the future.

To pre-empt the loss of their ecological legacy when they eventually sold their properties, landholders were considering novel ways to pass on the management knowledge they had acquired over time. Ken, a non-participant who had future plans for a covenant, spoke of 'progression planning not succession planning' when selling his land. Acutely aware of the experiential knowledge he had accrued over the space of a decade, Ken suggested he would favour a buyer who would be prepared to have him work (at no cost) as a property manager for a while, enabling knowledge exchange to take place through this process. As he mentioned, 'I've learnt so much about this place and it would be good to pass it on'.

This highlights a disconnect when it comes to *how* covenants protect legacies into the future. The legal protection offered through the addition of a covenant to a property title is only one part of the picture. The other is the continuity of conservation practices. The covenant as a mechanism does not offer a clear means for the landholder to pass on what has been learned through the human-environment relations that constitute conservation practice. This disconnect between passing on a legal agreement but not passing on the relevant conservation practices that support what the legal agreement ostensibly protects has been noted by some policy-makers. Evan, a retired bureaucrat who was previously involved in Australian federal environment policy noted:

> covenants in many cases... were pretty prescriptive. "Thou shalt, and if you don't, you..." which didn't really acknowledge what you're also buying... knowledge of country. And in fact, towards the end I was sort of saying, "We need a slightly different approach; we need a network of [landholders] who can demonstrate on their land what they're doing in relation to conservation and for them to become champions of private land conservation... we need to value and respect their knowledge.

As a result of this tension noted above, the protection of a covenant as a legally binding constraint on land use brought a great sense of relief, but also a strong self-awareness that changing ownership could disrupt legacy preservation. This left some landholders looking for alternative ways of passing on the kinds of 'knowledge of country' that Evan mentions above. For Ken, this meant an offer of free labour to help a new owner get accustomed to the conservation work that he had navigated over time. This might be substantive advice like what species of plant seems to grow well in what location, or even simple day-to-day knowledge, like which parts of the property become too wet for a tractor in the wintertime.

5 Conclusion

Beside the disconnect around legacy preservation, covenant objectives accorded closely with the processes that informed conservation practices in earlier chapters. Covenants were adopted by participants with both active and passive stewardship dispositions. While the scheme itself did not appear to challenge existing dispositions, different elements of the scheme may have been more appealing to different dispositions. For example, those with active dispositions were primarily the ones who wanted to protect the stewardship effort they had put in over time. On the other hand, Janet and Rob (passive dispositions) were more conscious of protecting their patch from clearance by future owners. There was no indication that adopting the scheme resulted in shifting these existing dispositions.

The experiences of landholders show that the security offered by covenants is highly sought after and a substantial relief in terms of securing conservation benefits that can extend beyond their property tenure. What is also of note, which is echoed in some of the literature on covenants, is that covenants are much more attuned to restricting private land uses through forgoing property rights, than they are about fostering conservation practice (Rissman, Owley, Shaw, & Thompson, 2015). The covenant logic works well for eliminating threats like development, but on its own, it is not enough to ensure an ongoing process of conservation. Land trusts acknowledge this as well, with the turnover of properties with covenants considered a potential risk to the security of the ecologies that covenants protect (Hardy et al., 2017). But the idea of 'fit it and forget it' (Rissman et al., 2013, p. 286) when it comes to signing up landholders to covenants does appear to be changing, with landholders being increasingly offered forms of practical support by the land trusts who sign them up. This is seen through Trust for Nature's stewardship scheme, which aims to get Trust officers out to properties to talk to covenant holders at least every few years.

It is clear that conservation practices need to be ongoing, regardless of land tenure, and that legal prescriptions alone will not do the work required for conservation on their own. However, landholder stories help to show that attention to conservation practice must be very specific; it matters 'how' conservation practice is understood and translated into an ongoing trajectory of conservation practice. This is partly about understanding landholder motives, but just as significantly, about understanding that conservation practice is co-produced, flexible but durable (especially over time) and bound up with other experiences of land use and interaction with ecologies. As we have shown, conservation practice is place specific, and place matters for linking the legal and practice dimensions of covenants. There is an evident need to find ways of transferring and translating stories of conservation practice, of directly sharing practices and then working collaboratively with other covenantors to address their challenges. The difficulties of undertaking conservation work in novel ecologies that are increasingly influenced by climate change only exacerbates the need for experiential, placed-based sharing of conservation practices. We discuss some possibilities around conservation practice-sharing in Chapter 7.

The covenant concept aligns closely with the value placed on stewarding the amenity property as private and 'owned' space. The sense that property rights ascribe a particular autonomy of land use appeared to prevent some landholders with covenants from suggesting what their neighbours might do with their land in terms of specific conservation practices. Yet, they were willing to talk broadly about the benefits of a covenant to their neighbours, in a manner that did not suggest that conservation was an obligation for rural-amenity property owners. On the other hand, despite covenants being grounded at the property scale in terms of the space they protect, some landholders understood them as discrete contributions to collective conservation benefits at a larger scale. This shows how private property could be framed in public good terms, reflecting the importance of a nuanced view of how private property shapes conservation practices (Massey, 1980).

The recognition of the disjuncture between landscape legacy and the legal protections offered by covenants, and the patchy efforts by landholders to amplify the benefits of their property-centric conservation efforts across the landscape, offer revealing insights into the performance of covenants as an environmental governance tool. As noted earlier, the retreat of the state in protected area conservation has furthered the abstraction of conservation to the scale of individual properties. And while landholders often seek large scale and temporally extended conservation benefits through their covenanting, the individual private property parcel presents as a challenging scale at which to achieve such outcomes. This is not to reify public protected areas as panacea, or to suggest private conservation through covenants is not an important contribution. What it does highlight is that private property and covenant arrangements have constraints in terms of the spatial reach and temporal continuity of conservation practices, and that landholders themselves can come to realise these constraints over the course of their land tenure. In thinking through ways of reducing these implications, conservation practice as a human-environment relation needs to be woven into our understanding of how covenants work as a governance mechanism.

Following on from this chapter, we continue our focus on neoliberal conservation initiatives by examining the use of market-based instruments for private land conservation. As with our review of conservation covenants, we seek to understand how rural-amenity migrants are co-opting such initiatives in ways that depart from the logic of the program in question.

References

Adams, W. M., Hodge, I. D., & Sandbrook, L. (2014). New spaces for nature: The re-territorialisation of biodiversity conservation under neoliberalism in the UK. *Transactions of the Institute of British Geographers, 39*(4), 574–588. https://doi.org/10.1111/tran.12050.

Bingham, H., Fitzsimons, J. A., Redford, K. H., Mitchell, B. A., Bezaury-Creel, J., & Cumming, T. L. (2017). Privately protected areas: Advances and challenges in guidance, policy and documentation. *Parks, 23*(1), 13–28. http://doi.org/10.2305/IUCN.CH.2017.PARKS-23-1HB.en.

Farmer, J. R., Knapp, D., Meretsky, V. J., Chancellor, C., & Fischer, B. C. (2011). Motivations influencing the adoption of conservation easements. *Conservation biology: The Journal of the Society for Conservation Biology, 25*(4), 827–834. https://doi.org/10.1111/j.1523-1739.2011.01686.x.

Fitzsimons, J. A., & Carr, C. B. (2014). Conservation covenants on private land: Issues with measuring and achieving biodiversity outcomes in Australia. *Environmental Management, 54*(3), 606–616.

Fletcher, R., & Breitling, J. (2012). Market mechanism or subsidy in disguise? Governing payment for environmental services in Costa Rica. *Geoforum, 43*(3), 402–411. https://doi.org/10.1016/j.geoforum.2011.11.008.

Fletcher, R., & Büscher, B. (2017). The PES conceit: Revisiting the relationship between payments for environmental services and neoliberal conservation. *Ecological Economics, 132*, 224–231. https://doi.org/10.1016/j.ecolecon.2016.11.002.

Gibson-Graham, J. K., Cameron, J., & Healy, S. (2013). *Take back the economy: An ethical guide for transforming our communities*. Minneapolis: University of Minnesota Press.

Hardy, M. J., Fitzsimons, J. A., Bekessy, S. A., & Gordon, A. (2017). Exploring the permanence of conservation covenants. *Conservation Letters, 10*(2), 221–230. https://doi.org/10.1111/conl.12243.

Kay, K. (2016). Breaking the bundle of rights: Conservation easements and the legal geographies of individuating nature. *Environment and Planning A, 48*(3), 504–522. https://doi.org/10.1177/0308518X15609318.

Klepeis, P., Gill, N., & Chisholm, L. (2009). Emerging amenity landscapes: Invasive weeds and land subdivision in rural Australia. *Land Use Policy, 26*(2), 380–392. https://doi.org/10.1016/j.landusepol.2008.04.006.

Lai, P. H., & Kreuter, U. P. (2011). Examining the direct and indirect effects of environmental change and place attachment on land management decisions in the Hill Country of Texas, USA. *Landscape and Urban Planning, 104*(3–4), 320–328. https://doi.org/10.1016/j.landurbplan.2011.11.007.

Lockie, S., & Higgins, V. (2007). Roll-out neoliberalism and hybrid practices of regulation in Australian agri-environmental governance. *Journal of Rural Studies, 23*(1), 1–11. https://doi.org/10.1016/j.jrurstud.2006.09.011.

Lockwood, M., & Davidson, J. (2010). Environmental governance and the hybrid regime of Australian natural resource management. *Geoforum, 41*(3), 388–398. https://doi.org/10.1016/j.geoforum.2009.12.001.

Macnaghten, P. (2008). Embodying the environment in everyday life practices. *The Sociological Review, 51*(1), 63–84. https://doi.org/10.1111/1467-954X.00408.

Mansfield, B. (2009). Property and the remaking of nature-society relations. In B. Mansfield (Ed.), *Privatization: Property and the remaking of nature-society relations* (pp. 1–13). Malden: Blackwell.

Massey, D. (1980). The pattern of landownership and its implications for policy. *Built Environment, 6*(4), 263–271. Retrieved from http://www.jstor.org/stable/23284724.

Moon, K., & Cocklin, C. (2011). A landholder-based approach to the design of private-land conservation programs. *Conservation Biology: The Journal of the Society for Conservation Biology, 25*(3), 493–503. https://doi.org/10.1111/j.1523-1739.2010.01639.x.

Morris, A. W. (2008). Easing conservation? Conservation easements, public accountability and neoliberalism. *Geoforum, 39*(3), 1215–1227. https://doi.org/10.1016/j.geoforum.2006.10.004.

Rissman, A. R., Bihari, M., Hamilton, C., Locke, C., Lowenstein, D., Motew, M., … Smail, R. (2013). Land management restrictions and options for change in perpetual conservation easements. *Environmental Management, 52*(1), 277–288. http://doi.org/10.1007/s00267-013-0091-z.

Rissman, A. R., Owley, J., Shaw, M. R., & Thompson, B. B. (2015). Adapting conservation easements to climate change. *Conservation Letters, 8*(1), 68–76. https://doi.org/10.1111/conl.12099.

Rissman, A. R., & Sayre, N. F. (2012). Conservation outcomes and social relations: A comparative study of private ranchland conservation easements. *Society*

& Natural Resources, 25(6), 523–538. https://doi.org/10.1080/08941920.2011.580419.

Selinske, M. J., Cooke, B., Torabi, N., Hardy, M. J., Knight, A. T., & Bekessy, S. A. (2016). Locating financial incentives among diverse motivations for long-term private land conservation. *Ecology and Society, 22*(2), 1–10.

Stuart, D., Benveniste, E., & Harris, L. M. (2014). Evaluating the use of an environmental assurance program to address pollution from United States cropland. *Land Use Policy, 39,* 34–43. https://doi.org/10.1016/j.landusepol.2014.03.009.

Urquhart, J., & Courtney, P. (2011). Seeing the owner behind the trees: A typology of small-scale private woodland owners in England. *Forest Policy and Economics, 13*(7), 535–544. https://doi.org/10.1016/j.forpol.2011.05.010.

Van Hecken, G., Bastiaensen, J., & Huybrechs, F. (2015). What's in a name? Epistemic perspectives and Payments for Ecosystem Services policies in Nicaragua. *Geoforum, 63,* 55–66. https://doi.org/10.1016/j.geoforum.2015.05.020.

CHAPTER 6

Market-Based Instruments and Conservation Practices on Private Land

Abstract The rise of market-based instruments (MBIs) as a conservation tool presents an important context in which to extend our understanding of how conservation practices interact with policy. This chapter centres on a reverse-auction MBI called 'EcoTender' in Victoria, Australia. We examine the types of conservation actions that emerge through landholders' participation in EcoTender and explore the way EcoTender is adopted, co-opted and resisted in different ways. This includes how landholders approach the costing of their own labour as part of the bidding process, the ways in which local ecologies may resist program prescriptions and the desire for social interaction amongst participants in a program that requires competition for cost efficiency. We conclude by reflecting on the future of MBIs for facilitating conservation practice.

Keywords Market-based instruments · Financial incentives · Conservation practice · More-than-human · Governance

1 Introduction

As the calls for more investment and policy development around private land conservation continue to grow, some recent scholarship has turned a more critical eye to the environmental governance contexts in which private land conservation sits, and the implications of these governance

B. Cooke and R. Lane, *Making Ecologies on Private Land*,
https://doi.org/10.1007/978-3-030-31218-3_6

arrangements for the way that conservation plays out in practice (Selinske et al., 2016). Central to these concerns are the consequences of neoliberal governance for transparency, equity and the effectiveness of public expenditure for private conservation (Morris, 2008; Rissman, Owley, Roe, Morris, & Wardropper, 2017). As noted in Chapter 4, unpacking the systematic assumptions of neoliberal governance critiques is necessary work. One of the vital things this book illustrates is that governance mentalities are rarely all-encompassing, meaning that people will not act entirely in accordance with the rationality of policies or programs when they engage with them (van Hecken, Bastiaensen, & Windey, 2015).

The recent rise of market-based instruments (MBIs) as a conservation tool presents an important yet under-examined context in which to extend an analysis of conservation practice. Existing research into the practice of MBIs has centred mainly on payments for ecosystem services (PES) in developing countries (see McElwee, Nghiem, Le, Vu, & Tran, 2014; Pattanayak, Wunder, & Ferraro, 2010; Sattler & Matzdorf, 2013), and the capacity of payments for ecosystem services (PES) to affect ongoing land-use change (van Hecken & Bastiaensen, 2010). Here we build on this work by examining how MBIs that use a reverse-auction tender model for private land conservation in a western capitalist context are materialising in Victoria, Australia.

In keeping with our focus on conservation practice, we are interested in how relationships between people and ecologies actively shape the way policy is enacted. As with the analysis of conservation covenants in the previous chapter, our specific focus is on how conservation practices can disrupt the logic of MBIs (McElwee et al., 2014; van Hecken et al., 2015). Besides landholders themselves, local ecologies, and plants in particular, may resist the logic of MBIs through their growth, spread, change over time and entanglements with human agency. Unpacking the way that ecologies conform or resist being enrolled as policy subjects is particularly important for understanding the capacity for conservation policy to respond to environmental change (Head et al., 2015). This is partly what drives the increasing attention to nonhuman agency in conservation.

This chapter centres on a reverse-auction MBI called 'EcoTender', which 18 of our study participants were involved in. We examine the types of conservation actions that emerged through their participation in EcoTender and explore the tensions between the reverse-auction tender model and landholder aspirations for conservation practice. This includes how landholders approach the costing of their own labour as part of the bidding process, the ways in which local ecologies may resist program prescriptions framed around discrete private property parcels and the desire for social interaction amongst participants in a program that requires competition to drive down costs. We conclude by reflecting on the need and role for financial incentives for conservation practice.

2 The Interaction of MBIs and Conservation Practice

While much of the research on neoliberal conservation has centred on structural critiques (Schwartz, 2013), a growing body of work has begun to show the divergence that can occur between the intentions of neoliberal conservation and the on-ground outcomes (McElwee et al., 2014; Roth & Dressler, 2012; van Hecken et al., 2015; Vatn, 2015). As noted in the previous chapter, in the context of western capitalist systems of governance, the analysis of MBIs for PLC in practice remains a largely understudied component of conservation governance. We are interested in addressing this gap by exploring how and why individual landholders and ecologies can co-opt, resist or re-appropriate MBIs through conservation practice. As we will explore, MBIs can carry embedded assumptions about the composition, function and flourishing of ecologies, which are often heavily contingent on socio-ecological context.

For private land conservation, the discourses of efficiency and flexibility have helped to legitimise market-based mechanisms. Approaches to environmental governance that reflect the logic of privatisation and marketisation are touted as simultaneously strengthening economic and ecological sustainability and increasing both policy efficiency and flexibility (Fletcher, Dressler, Büscher, & Anderson, 2016). Efforts to generate social and ecological benefits through MBIs highlight the competing logics that can be inherent in this form of governance (Owley & Rissman, 2016; Wynne-Jones, 2012). While MBIs contribute to nature commodification, at the same time they aim to promote social learning, community empowerment and the protection of threatened species (Higgins, Dibden, & Cocklin, 2012; Zammit, 2013). Yet, more work is needed to understand how these 'hybrid' but also potentially contradictory forms of governance that are blending market logics with an orientation for public benefits actually intersect with conservation practice on private land.

We focus on landholder participation in 'EcoTender', operated by the state government of Victoria, Australia. EcoTender and its partner scheme,

'BushTender', were active between 2001 and 2016, and represent some of the earliest trials of reverse-auction MBIs for conservation (Whitten, Reeson, Windle, & Rolfe, 2013; Zammit, 2013). Both schemes were part of a broader 'EcoMarkets' initiative that aimed to provide incentives to private landholders to manage land for conservation through the use market mechanisms (Blackmore & Doole, 2013; Stoneham, Chaudhri, Ha, & Strappazzon, 2003; Whitten et al., 2013). These markets operated with the Victorian Government serving as the buyer of environmental goods in the interests of the public. EcoTender saw landholders in a specified region competing for a finite pool of government funds through a blind auction. Landholders in this region were invited to submit bids for the cost of conservation work they sought to undertake on their property. Like all reverse auctions, the process was designed to increase cost efficiency by creating competition for funds between landholders (Stoneham et al., 2003). Bids that represented the best value for money in terms of the conservation benefits generated (according to an Environmental Benefits Index developed by the government) would then be funded (Hajkowicz, 2009). After the EcoTender contracts examined here were concluded, no new tenders were opened under this Victorian Government scheme. Conservation tenders like this are now primarily operated by regional governance institutions in Victoria that operate at the catchment or watershed scale.

Reverse-auction tenders like EcoTender are considered an example of a payment approach to MBIs for conservation (Cooke & Moon, 2015; Pirard, 2012). The existence of competition between landholders for funds, the way landholder costs and ecological benefits are revealed through the bidding process, and the framing of the state as a buyer of conservation benefits are the primary MBI characteristics of EcoTender (Pirard, 2012; Wunder, 2015; Wunder, Engel, & Pagiola, 2008). These characteristics, in combination with the emphasis and reliance on private land rights for program delivery, reveal an assumption that market logics offer the solution to managing the problems of environmental degradation (Fletcher & Büscher, 2017). It is the facilitation of "market-oriented arrangements" (Pellizzoni, 2011, p. 796) by the state as a means for addressing environmental issues that reveals the neoliberal rationality of EcoTender.

The conservation actions supported by EcoTender focused on ecological restoration to pre-colonial benchmarks, the eradication of invasive weeds and the management and protection of existing patches of native vegetation (including fencing these patches). Since the early 2000s, there has been widespread adoption of tender schemes in Australia and growing interest and implementation globally (Whitten et al., 2013; Zammit, 2013), hence the importance of exploring the nuances of their operation.

2.1 Landholder Participant Forum

In addition to the research methods already noted in the introduction to this book, landholders who were involved in EcoTender during the 2016 fieldwork were also invited to a participant forum. The landholder participant forum was an opportunity for participants to gather together with Benjamin at the end of the research process to discuss findings and talk with each other about their EcoTender experience. There were 24 participants in the forum, including some local catchment/watershed management officers. The forum was open to all EcoTender participants from the West Gippsland trial, whether they were interviewed for the research or not. The engagement with landholders and their interactions during the forum were part of the research process, with particular emphasis on how the discussions facilitated between participants reinforced the importance of social networks for conservation.

The forum provided an opportunity for the researcher to present the themes discussed in the next section back to participants, allowing the forum to serve as a triangulation method for checking whether the themes identified by the researcher were those that participants themselves considered central to their experience. While participants did not substantively challenge the themes, there were some revisions made as a result of participant feedback, which were fed back into the analysis process.

3 Tender Participation as Conservation Practice

In this section, we explore three themes of significance for how conservation tenders operate in practice: uncosted labour in bids, emergent novel ecosystems and the absence of social networks. We place a strong emphasis on the creative and unanticipated co-option of payments, given their prominence, and the implications this raises for using MBIs in PLC. However, we also highlight the alignments that occur between landholder practice and the logic of conservation tenders, to ensure a comprehensive and representative engagement with the empirical material.

The idea that some conservation work should exist outside the 'service contract' that had been established through the tender process proved to be a unique insight into the way environmental stewardship was perceived and practiced. Of the 18 research participants with an EcoTender contract, only five included the cost of their own labour as part of their bid. There was a strong sense that the labouring that constituted conservation practice was an altruistic contribution that participants should provide. As Ingrid noted, "(The labour is) my part, you've got to put in yourself".

For landholders who already possess a strong conservation ethic (this covered a large proportion of our study participants, given their pre-existing conservation efforts discussed earlier), the money associated with EcoTender provided a unique opportunity to implement existing plans. However, the desire to be an active steward and contribute altruistically to conservation meant most landholders sought to volunteer their labour. As Adam identified, "at the end we're only custodians of the land", and the goal of conservation must be to improve the condition of land over your tenure. The desire to be an environmental steward and to leave a positive legacy is a well-established motivation for landholders involved in PLC (Lokocz, Ryan, & Sadler, 2011; Selinske, Coetzee, Purnell, & Knight, 2015):

> Jacinta I put all the labour in (for rigging) the fences and planting the trees, so it was really we were looking for the cost of the materials and the cost of the trees.
>
> Benjamin (First Author): Did you cost your labour?
>
> Jacinta: No. No… just materials. Because when we bought the property, I wanted to do this, before I even knew what an Ecotender was… we thought, wouldn't that be great revegetating, just as a project to do for ourselves sort of thing.

This reveals the way that conservation practice is bound up with other expressions and experiences of land use and ownership, which might align more with intrinsic, recreational or even aesthetic motivations for doing conservation work. The difference here is that this conservation work is being done under contract, which changes the dynamic and the obligation for delivery. Kim and Terry noted this, saying "it will be nice to just plug along at my own pace and not be subject to a timeframe" now that their contract is complete.

The landholder forum conducted at the conclusion of the fieldwork emphasised the uncertainty that many landholders had about how the process of pricing their bid. Terms like "flying blind" and "guess work" were used on a number of occasions to emphasise the level of uncertainty about what might constitute a competitive bid. Moreover, landholders feared missing out on getting at least some funding for conservation practices they were already undertaking, even if it did not cover the total cost of the works proposed in their bid. Across the participant forum and the interviews, three landholders also revealed that they thought costing your labour was not allowed. While this bodes well for the program in terms of reducing costs, it also reveals a risk to the maintenance of conservation benefits in the longer term. As Janet noted, the only question about whether their conservation efforts would continue beyond their contract would be "how long are we going to be able to physically manage this property?" Most participants did not cost the labour associated with weed management, meaning there is a risk to achieving this objective if landholders become physically incapable of doing the work.

As Jacinta noted, it was very difficult "trying to keep up with things" when you are volunteering your labour and the work was unfamiliar. Experiences like this saw a number of landholders reflect on the fact that "it would be good to get some help with (labour)" (Janet) if they were to apply for funding in the future. One landholding couple who did cost their labour did so at low rates to reduce costs, but also because they considered themselves to be lay conservation practitioners:

> Kim and Terry: [we costed labour at] "about $40 per tree, because we're just rank amateurs, we're not worth that much."

There were, however, four landholders that sought to cost their bids accurately. For two of these participants, including a "realistic" (Colin) costing was important, as they did not have the financial capacity to undertake the work if the full costs were not reflected in the bid. While all of these realistic bidders found it "very difficult to judge" (Faye and Michael) what their costs would be, they did not want to be burdened with having to fund too much of their project themselves.

The region in which EcoTender operated has a history of intensive landscape modification for agriculture following colonial invasion. While First Nations Peoples actively shaped this landscape, the large-scale clearing and rapid intensification of grazing, cropping and dairy post-invasion are the reason why ecological restoration efforts are being pursued. However, changes to soil composition, introduced flora and fauna, and climate change mean that traditional efforts to bring back benchmark pre-colonial ecologies have been challenging, if not undesirable, given future climate scenarios and major environmental change (Mansfield et al., 2014). EcoTender restoration was largely informed by a combination of historical benchmark ecologies, with some flexibility for amendments on a case-by-case basis. While restoration was a common component of landholder bids, many landholders encountered difficulties restoring landscapes to historic benchmarks. Through their experiences, we see the way that ecologies—and we focus on plants in particular—can be difficult to enrol as obedient participants in conservation governance. Ecologies can be unpredictable and unruly subjects and 'act back' when it comes to policy interventions.

Many landholders dealt with the challenge of managing plantings that had not materialised in the way they had intended. All but two participants had a restoration component to their conservation work, so the ecologies that were emerging possessed a high degree of novelty. The types of novel outcomes included a failure of seedlings to flourish, the dieback of new plantings and established plants around them, secondary infestations of weed species that had been cleared prior to planting, environmental change factors like drought or flooding and the loss of plantings from grazing by native and introduced fauna. Another common experience was the spread of plants from adjoining properties through seeds blown on the wind or washed over in floods.

The ubiquity of eucalypt species in Victoria means there were few restoration efforts that did not include eucalypts in the approved Ecological Vegetation Class (EVC) list of plantings. Yet, there are many regions in Victoria where eucalypts are not considered to be part of the pre-colonial ecology. This was the case for Ahmed, who did not include eucalypts in his planting. However, there were sections of property with an adjoining public land reserve that was a different EVC, which did include eucalypts. Subsequent to his EcoTender planting, Ahmed observed eucalypts from public land that adjoined his property starting to sprout in this patch, eventually becoming the most dominant species in the planted area. The rate of growth and substantial number of recruiting species led to a belief on his part that eucalypts did indeed "belong" in this highly modified landscape. Indeed, six other participants noted the growth of native species that they did not plant themselves within their revegetation areas.

Two participants also wondered whether what they were recreating an ecology that did not ultimately reflect what existed there prior to colonial invasion. Both participants were conscious that the bush had been cultivated by First Nations Peoples prior to European invasion, and they wondered whether what was being reproduced was a static ecology without human influence. As Devon explained:

> Compared to what [this land] would have looked like say pre-1770, you wouldn't have had that understorey as thick as it is. Your trees would have probably been thinner with more mature trees. Is that what we're supposed to be getting back to? So what are we protecting? That's something that's starting to make me sceptical [about what we are doing] again...

This experience reinforces that restoration sites are not blank canvases that exist in a spatial and temporal vacuum—the history of land use and the ecologies on surrounding lands will play a substantial role in how restoration unfolds. With the potential for historical ecological benchmarks to become an even more difficult—and potentially problematic—restoration target into the future, there is a need to consider what we aim to achieve through restoration work. It may also mean considering ways in which conservation programs could be decoupled from private property as a scale of emphasis (Cooke & Moon, 2015). While some participants took a hands-off approach to managing unexpected plant growth in their restoration site, commenting that "it will work itself out" (Elise), others felt a level of anxiety that the unexpected plant growth could constitute a breach of their EcoTender contract.

The planting process also generated habitat interactions that participants did not anticipate as being an issue at the outset. A notable example is the impacts that fauna grazing and habitat creation could have on the health and survivability of the species planted. Often, the grazing species were native animals (such as wallabies or wombats), and landholders developed a negative perception of these animals as they had undone all their hard work to establish the planting. As Jacinta noted, "if it weren't for the bloody Wombats" grazing on her new plantings then she would not have had to spend additional money from her own pocket to replace some of the species lost. When surrounding ecologies are fragmented and resources are scarce, grazing fauna may be more likely to target new growth in restoration areas (depicted in Fig. 1).

When native plants spread across property boundaries, they can call attention to the range, distribution and reproduction of that species in ways that reveal wider ecological function. As in the example above, the presence and movement of species were powerful catalysts for triggering awareness that the property scale was not always suited to conservation

Fig. 1 Jacinta noted that resident wombats in the forested segment of the adjacent property (top of image) had come onto her property to eat the planted seedlings (centre of image, upslope of the fence)

efforts. As Wendy noted when looking across the district from a high point on the property, "on a windy day you can see the seed heads blowing across the hill". Movement of weeds across properties challenged the contractual obligation through EcoTender to keep restoration plantings weed-free, which made many acutely aware that the fortunes of their property cannot be separated from what goes on in the wider landscape.

The downside to a model that requires competition between bidding landholders to drive down costs is that participants can go the duration of their contract without encountering other participants. EcoTender has no direct mechanism for gathering people together before, during or after bids are awarded. This is in contrast to initiatives like Landcare or Conservation Management Networks (discussed further in Chapter 7). While landholders were monitoring basic outcomes of their conservation practices for reporting to DELWP, this data was not shared amongst participants, and participants were not sure what DELWP did with their monitoring data. A key criticism of auction-style programs, therefore, is that they can fail to build collaborative, cross-boundary conservation efforts. Kevin's property was on a creek, and he felt that the program would have been better served encouraging collaboration amongst creek-frontage properties:

> To my mind it would have been better if the government took a more holistic approach and looked at the whole creek and came along and said, "Okay, we need to revegetate that creek and make grants available to people along a particular creek so that it was one whole area rather than just one bit in the middle in isolation."

Beyond on-ground collaboration, eight landholders spoke unprompted about the fact that they would like to meet other participants and have a participant network. The central reason for wanting social connection was to gain some reassurance that they were not the only ones encountering challenges, difficulties or unexpected outcomes from their activities (plants not growing, weeds returning, for example). Kim and Terry captured this feeling when describing the fact that it would be nice to have a sense of "shared experience" in EcoTender. In terms of shared experience, landholders spoke about being interested in knowing what others had been doing as part of their contract, and that specific advice or support was not really necessary. For some landholders, this point was related to the fact that the EcoTender contract created a sense of obligation that they had not felt with their prior conservation work.

While it could be argued that EcoTender need not concern itself with enabling social interaction and learning when it can happen outside of the program, there are a host of issues that arise through program participation and the specifics of EcoTender operation that participants would like to be able to discuss with one another. The ability to work through issues or uncertainties surrounding PLC programs (Moon, Marshall, & Cocklin, 2012) and tender programs in particular (Blackmore & Doole, 2013) has been previously reported by landholders. Moreover, not all participants enjoy neighbours who are undertaking conservation practices, and this can leave some landholders isolated.

Despite the general view that a network would be beneficial, five landholders struggled to think of any particular benefits that might come from interaction with other participants. One landholder wondered, "what's there to talk about?" (Ingrid) concerning conservation work. Conservation programs cannot assume all participants will see value in knowledge sharing, as some are happy working individually on their own property. This appears especially true in rural-amenity landscapes, where new in-migrants can be more inwardly focused on managing their own property without wider interaction and engagement (Gill, Klepeis, & Chisholm, 2010; Yung & Belsky, 2007). For participants like this, the property-centric, individual landholder engagement model suited their approach to conservation.

4 Conclusion

The stories landholders told about their adoption of the EcoTender program bring to light the need for the prioritisation of local socio-ecological dynamics in the design and implementation of market instruments (Selinske et al., 2016). What must now be explicitly acknowledged is the need to challenge PLC policy that uncritically adopts MBIs because they are fashionable, without considering their alignment to the needs and circumstances of people and ecologies on the ground (Sorice & Donlan, 2015). The way many EcoTender participants struggled to conceptualise their stewardship labour as a service to be costed and delivered to a contractual deadline presented a unique insight into the mismatch between policy logic and landholder practice; the real consequences of this mismatch being uncertainty and anxiety on the part of landholders about their capacity to fund and fulfil their conservation contracts. To ignore this context is to render invisible the diverse socio-ecological dynamics of conservation practice and the diverse experiences of those dynamics amongst policy-makers, researchers and landholders.

MBIs for private land conservation must explicitly acknowledge that ecologies act and react to human intervention through conservation practices in ways that are not uniform, easily anticipated or isolated from surrounding socio-ecological processes (Head et al., 2015; Ogden et al., 2013). When MBIs engage people through private property, the mobility of plants and fauna (both wanted and unwanted) at scales above the property parcel are not reflected. As this study demonstrated, wider ecosystem interactions drew people into landscape-scale conservation challenges that EcoTender had no formal mechanism for addressing. In this way, ecologies pushed back against efforts to control conservation practice at the scale of the property parcel. This suggests the need for coordinated implementation at larger scales. Constraining PLC policy to individual contracts at the scale of individual properties risks closing down opportunities for collective conservation at a time when landscape-scale responses are deemed increasingly necessary (Fitzsimons, Pulsford, & Wescott, 2013).

The experiences of landholders in this study raise questions about the rigidity of MBI contracts if we are anticipating uncertainty in the way ecologies respond to the types of interventions pursued through conservation practice. This is especially true when we consider the knock-on effects that are created by the initial conservation work undertaken by landholders (such as triggering secondary weed infestations) that will have impacts well beyond discrete contracts. MBIs such as reverse-auctions also require landholders to commit to a program of works at the outset given the need to formulate a bid based on proposed works, restricting their capacity to respond to unanticipated outcomes along the way. As finite interventions, programs like EcoTender that often run for short periods (5–10 years) need to accommodate the history of landscape modification and the contingent ways people and ecologies are responding to change. The tensions between short-term contracts and ongoing conservation practice suggest payments must be part of a wider PLC response in which landholders have a stronger role in policy design, rather than being the lynchpin around which those programs operate (Muradian et al., 2013; Selinske et al., 2016). For example, landholders could play an important role in negotiating the make-up of species types that are used in restoration by drawing on their experiential learning, landscape observations and landscape legacy, while still ensuring that ecological function, diversity and abundance are enhanced by the restoration process (Collard, Dempsey, & Sundberg, 2014).

In many cases, conservation initiatives involving MBIs require private property rights as a precondition for competition between property owners for funding (Holmes & Cavanagh, 2016). Deploying market logics in this way implies a view that potential cost efficiency gains from competition between landholders for funding is preferable to collaboration between landholders around shared conservation goals. Bounding the private property as the focal point for private land conservation also risks further entraining conservation to the private property parcel at a time of growing advocacy for landscape-scale conservation that acknowledges the flows and networks of multispecies assemblages (see Laven, Jewiss, & Mitchell, 2013, for example). More work is needed to imagine ways in which individual property rights can be decentred as the cornerstone of conservation funding models.

The manner in which landholders engaged with EcoTender was shaped by their existing conservation practice over time. These experiences did not always align with the logic of conservation tenders. Indeed, many of the stories explored above show that people used EcoTender to leverage conservation efforts in ways that contradicted the intent of a program centred on financial incentives. Their experiences also show that ecologies are not easily entrained to the property-centric conservation efforts that often accompany MBIs like EcoTender. We are prompted to question the appropriateness of conservation tenders in this context and to consider whether they could be reoriented to support conservation as the kind of relational practice that we articulate in this book. At a broader level, we question whether MBIs as policy tools may have limited capacity to respond to the emergent challenges presented by PLC, especially in dynamic rural-amenity landscapes.

In the final chapter of the book, we reflect on how attention to conservation practice might steer us to think in more subtle and diverse ways when it comes to private land conservation in rural-amenity landscapes. We build on our exploration of conservation practice, nonhuman agency and interactions with policy interventions like covenants and MBIs, to explore wider possibilities for private land conservation in rural-amenity landscapes.

References

Blackmore, L., & Doole, G. J. (2013, November). Drivers of landholder participation in tender programs for Australian biodiversity conservation. *Environmental Science & Policy, 33*, 143–153. https://doi.org/10.1016/j.envsci.2013.05.010.

Collard, R. C., Dempsey, J., & Sundberg, J. (2014). A manifesto for abundant futures. *Annals of the Association of American Geographers, 105*(2), 322–330. https://doi.org/10.1080/00045608.2014.973007.

Cooke, B., & Moon, K. (2015). Aligning 'public good' environmental stewardship with the landscape-scale: Adapting MBIs for private land conservation policy. *Ecological Economics, 114*, 152–158. https://doi.org/10.1016/j.ecolecon.2015.03.027.

Fitzsimons, J., Pulsford, I., & Wescott, G. (2013). Linking Australia's landscapes: An introduction. In J. Fitzsimons, I. Pulsford, & G. Wescott (Eds.), *Linking Australia's landscapes: Lessons and opportunities from large-scale conservation networks* (pp. 3–6). Melbourne: CSIRO Publishing.

Fletcher, R., & Büscher, B. (2017). The PES conceit: Revisiting the relationship between payments for environmental services and neoliberal conservation. *Ecological Economics, 132*, 224–231. https://doi.org/10.1016/j.ecolecon.2016.11.002.

Fletcher, R., Dressler, W., Büscher, B., & Anderson, Z. (2016). Questioning REDD+ and the future of market-based conservation. *Conservation Biology*. https://doi.org/10.1017/CBO9781107415324.004.

Gill, N., Klepeis, P., & Chisholm, L. (2010). Stewardship among lifestyle oriented rural landowners. *Journal of Environmental Planning and Management, 53*(3), 317–334. https://doi.org/10.1080/09640561003612890.

Hajkowicz, S. (2009). The evolution of Australia's natural resource management programs: Towards improved targeting and evaluation of investments. *Land Use Policy, 26*(2), 471–478. https://doi.org/10.1016/j.landusepol.2008.06.004.

Head, L., Larson, B., Hobbs, R. L., Atchison, J., Gill, N., Kull, C., & Rangan, H. (2015). Living with invasive plants in the anthropocene: The importance of understanding practice and experience. *Conservation and Society, 13*(3). https://doi.org/10.4103/0972-4923.170411.

Higgins, V., Dibden, J., & Cocklin, C. (2012). Market instruments and the neoliberalisation of land management in rural Australia. *Geoforum, 43*(3), 377–386. https://doi.org/10.1016/j.geoforum.2010.10.002.

Holmes, G., & Cavanagh, C. J. (2016). A review of the social impacts of neoliberal conservation: Formations, inequalities, contestations. *Geoforum, 75*, 199–209. https://doi.org/10.1016/j.geoforum.2016.07.014.

Laven, D. N., Jewiss, J. L., & Mitchell, N. J. (2013). Toward landscape-scale stewardship and development: A theoretical framework of United States National Heritage Areas. *Society & Natural Resources, 26*(7), 762–777.

Lokocz, E., Ryan, R. L., & Sadler, A. J. (2011). Motivations for land protection and stewardship: Exploring place attachment and rural landscape character in Massachusetts. *Landscape and Urban Planning, 99*(2), 65–76. https://doi.org/10.1016/j.landurbplan.2010.08.015.

Mansfield, B., Biermann, C., McSweeney, K., Law, J., Gallemore, C., Horner, L., & Munroe, D. K. (2014). Environmental politics after nature: Conflicting socioecological futures. *Annals of the Association of American Geographers, 105*(2), 284–293. https://doi.org/10.1080/00045608.2014.973802.

McElwee, P., Nghiem, T., Le, H., Vu, H., & Tran, N. (2014). Payments for environmental services and contested neoliberalisation in developing countries: A case study from Vietnam. *Journal of Rural Studies, 36*, 423–440. https://doi.org/10.1016/j.jrurstud.2014.08.003.

Moon, K., Marshall, N., & Cocklin, C. (2012). Personal circumstances and social characteristics as determinants of landholder participation in biodiversity conservation programs. *Journal of Environmental Management, 113*, 292–300. https://doi.org/10.1016/j.jenvman.2012.09.003.

Morris, A. W. (2008). Easing conservation? Conservation easements, public accountability and neoliberalism. *Geoforum, 39*(3), 1215–1227. https://doi.org/10.1016/j.geoforum.2006.10.004.

Muradian, R., Arsel, M., Pellegrini, L., Adaman, F., Aguilar, B., Agarwal, B., & Corbera, E. (2013). Payments for ecosystem services and the fatal attraction of win-win solutions. *Conservation Letters, 6*(4), 274–279. https://doi.org/10.1111/j.1755-263x.2012.00309.x.

Ogden, L., Heynen, N., Oslender, U., West, P., Kassam, K. A., & Robbins, P. (2013). Global assemblages, resilience, and earth stewardship in the anthropocene. *Frontiers in Ecology and the Environment, 11*(7), 341–347. https://doi.org/10.1890/120327.

Owley, J., & Rissman, A. R. (2016). Trends in private land conservation: Increasing complexity, shifting conservation purposes and allowable private land uses. *Land Use Policy, 51*, 76–84. https://doi.org/10.1016/j.landusepol.2015.10.026.

Pattanayak, S. K., Wunder, S., & Ferraro, P. J. (2010). Show me the money: Do payments supply environmental services in developing countries? *Review of Environmental Economics and Policy, 4*(2), 254–274. https://doi.org/10.1093/reep/req006.

Pellizzoni, L. (2011). Governing through disorder: Neoliberal environmental governance and social theory. *Global Environmental Change, 21*(3), 795–803. https://doi.org/10.1016/j.gloenvcha.2011.03.014.

Pirard, R. (2012). Market-based instruments for biodiversity and ecosystem services: A lexicon. *Environmental Science and Policy, 19–20*, 59–68. https://doi.org/10.1016/j.envsci.2012.02.001.

Rissman, A. R., Owley, J., Roe, A. W. L., Morris, A. W., & Wardropper, C. B. (2017). Public access to spatial data on private-land conservation. *Ecology and Society, 22*(2), 24.

Roth, R. J., & Dressler, W. (2012). Market-oriented conservation governance: The particularities of place. *Geoforum, 43*(3), 363–366. https://doi.org/10.1016/j.geoforum.2012.01.006.

Sattler, C., & Matzdorf, B. (2013). PES in a nutshell: From definitions and origins to PES in practice—Approaches, design process and innovative aspects. *Ecosystem Services, 6*, 2–11. https://doi.org/10.1016/j.ecoser.2013.09.009.

Schwartz, K. Z. S. (2013). Panther politics: Neoliberalizing nature in southwest Florida. *Environment and Planning A, 45*(10), 2323–2343. https://doi.org/10.1068/a45294.

Selinske, M. J., Coetzee, J., Purnell, K., & Knight, A. T. (2015). Understanding the motivations, satisfaction, and retention of landowners in private land conservation programs. *Conservation Letters, 8*(4), 282–289. https://doi.org/10.1111/conl.12154.

Selinske, M. J., Cooke, B., Torabi, N., Hardy, M. J., Knight, A. T., & Bekessy, S. A. (2016). Locating financial incentives among diverse motivations for long-term private land conservation. *Ecology and Society, 22*(2), 7.

Sorice, M. G., & Donlan, C. J. (2015). A human-centered framework for innovation in conservation incentive programs. *Ambio, 44*(8), 788–792. https://doi.org/10.1007/s13280-015-0650-z.

Stoneham, G., Chaudhri, V., Ha, A., & Strappazzon, L. (2003). Auctions for conservation contracts: An empirical examination of Victoria's BushTender trial. *The Australian Journal of Agricultural and Resource Economics, 47*(4), 477–500. https://doi.org/10.1111/j.1467-8489.2003.t01-1-00224.x.

van Hecken, G., & Bastiaensen, J. (2010). Payments for ecosystem services in Nicaragua: Do market-based approaches work? *Development and Change, 41*(3), 421–444.

van Hecken, G., Bastiaensen, J., & Windey, C. (2015). Towards a power-sensitive and socially-informed analysis of payments for ecosystem services (PES): Addressing the gaps in the current debate. *Ecological Economics, 120*, 117–125. https://doi.org/10.1016/j.ecolecon.2015.10.012.

Vatn, A. (2015). Markets in environmental governance—From theory to practice. *Ecological Economics, 117*(1), 225–233. https://doi.org/10.1016/j.ecolecon.2014.07.017.

Whitten, S. M., Reeson, A., Windle, J., & Rolfe, J. (2013, December). Designing conservation tenders to support landholder participation: A framework and case study assessment. *Ecosystem Services, 6*, 1–11. https://doi.org/10.1016/j.ecoser.2012.11.001.

Wunder, S. (2015). Revisiting the concept of payments for environmental services. *Ecological Economics, 117*, 234–243. https://doi.org/10.1016/j.ecolecon.2014.08.016.

Wunder, S., Engel, S., & Pagiola, S. (2008). Taking stock: A comparative analysis of payments for environmental services programs in developed and developing countries. *Ecological Economics, 65*(4), 834–852. https://doi.org/10.1016/j.ecolecon.2008.03.010.

Wynne-Jones, S. (2012). Negotiating neoliberalism: Conservationists' role in the development of payments for ecosystem services. *Geoforum, 43*(6), 1035–1044. https://doi.org/10.1016/j.geoforum.2012.07.008.

Yung, L., & Belsky, J. M. (2007). Private property rights and community goods: Negotiating landowner cooperation amid changing ownership on the Rocky Mountain front. *Society & Natural Resources, 20*(8), 689–703. https://doi.org/10.1080/08941920701216586.

Zammit, C. (2013, March). Landowners and conservation markets: Social benefits from two Australian government programs. *Land Use Policy, 31*, 11–16. https://doi.org/10.1016/j.landusepol.2012.01.011.

CHAPTER 7

Conclusion: Collective Conservation Practice in Rural-Amenity Landscapes

Abstract Having explored the ways that everyday conservation practices are constituted, the final chapter takes these insights and considers how we might advance conservation efforts in rural-amenity landscapes along two distinct but related lines. The first of these is to consider what could be done to build a more collective approach that acknowledges landscapes legacy through grass-roots initiatives and the reorientation of existing structures and institutional processes. The second line of reflection is to take seriously the need to re-imagine (western, liberal) private property relations that are more attuned to nonhuman agency and the unceded lands of First Nations Peoples in settler-colonial nations, to consider what other forms of property rights could be experimented with in rural-amenity landscapes.

Keywords Collective · Private land · Settler-colonial · Property · Conservation

To reflect on the possible futures for private land conservation in rural-amenity landscapes, we return to the conundrum faced by Liz at the beginning of this book: How should we be practicing conservation in light of disturbance legacies, contested knowledge and uncertain futures? Having explored the ways that everyday conservation practices are constituted throughout the book's chapters, our aim in this concluding chapter is to take these insights and consider how we might advance conservation efforts

B. Cooke and R. Lane, *Making Ecologies on Private Land*,
https://doi.org/10.1007/978-3-030-31218-3_7

in rural-amenity landscapes. The first of these is to propose how more collective approaches to conservation could be built that acknowledge landscape legacies through grass-roots initiatives and the reorientation of existing structures and institutional processes. The second line of reflection is to respond to the need to re-imagine (western, liberal) private property relations that are more attune to nonhuman agency and consider what other forms of property could be experimented with in rural-amenity landscapes.

To begin, it is worth reiterating that rural-amenity landscapes are highly significant for nature conservation. These landscapes often contain the last vestiges of ecologies that have been fragmented through a host of land uses and landscape modifications. In Australia, this includes being home to individual threatened species that once thrived in the regions that have been decimated by European agricultural practices and the ongoing expansion of settler cities. And because of the historic tendency for public protected areas to be established in places with little natural resource value, remnant ecologies in rural-amenity regions are over-represented on private land. Yet, ongoing fragmentation of landscapes and small properties (amongst other factors) present unique challenges. Rather than retreat from these challenges, rural-amenity landscapes offer a chance for deep reflection and experimentation when it comes to imagining private land conservation futures in the face of socio-ecological crisis.

1 Cultivating Collective Conservation Attuned to Dynamic Landscapes

In rural-amenity landscapes, we advocate for private land conservation goals based on collectively navigating uncertain trajectories rather than returning to historic ecological benchmarks. This requires that people share experiences of place-based conservation practices that they participate in. In conservation practices and the dispositions that can solidify around them, we have seen the particular ways that people come to know ecologies. This knowledge is a kind of bodily intelligence that emerges through intimate and ongoing interactions (Simpson, 2019). Efforts to engage landholders in a wider policy or political project for conservation across private land will need to engage with this situated experiential knowledge in order to gain traction and legitimacy. This means rejecting the idea of conservation practice as general, disembodied, human-centric and easily transferable from one site to the next.

There are many existing models for conservation that could be harnessed for sharing conservation practice and experience to enable more collective approaches across private land. In Australia (and elsewhere), Landcare offers a community-based model for action across different land tenures (Tennent & Lockie, 2012). Collective action through Catchment Management Authorities in Australia (at the catchment or watershed scale) also offer a model where ecologies might be conceived of at scales beyond the individual property parcel. Watershed governance more broadly has had some success in bringing different actors together for developing a shared vision and management agenda (Hill, Collins, & Vidon, 2018). An Australian model that has received less attention in the conservation literature is Conservation Management Networks (CMNs). These networks centre on particular biomes or remnant vegetation communities and then look to foster cooperation amongst human actors and organisations within the range of that vegetation community. For example, the Gippsland Plains CMN in Victoria centres on the Gippsland Redgum Grassy Woodland, bringing together public and private landholders in pursuit of conservation action for this nationally threatened ecology (Edwards & Fox, 2013). As with Landcare though, CMNs are not as prevalent in rural-amenity regions, as they are mainly organised around large, core areas of habitat on public land.

While in some regions, existing models may well be sufficient for sharing conservation practice and experiential knowledge, in others there is likely to be a need for cultivating new networks and collaborations. In either case, the CMN model that foregrounds particular ecologies in the ways landscape-scale collaborations are organised offers useful principles that connect with the types of practices explored in this book. Taking ecologies, rather than individual properties, as the focus around which human activity and endeavour is formulated is more sensitive to nonhuman agency. The CMN model also addresses the issue of vague or unfocused calls to collaboration that are detached from any particular socio-ecological interest or concern (Pratt, 2012). It can incorporate existing property-scale initiatives like nature covenants, by connecting them to a bigger picture for regional conservation across the network, which addresses concerns expressed by a number of landholders in Chapter 5. Yet, when it comes to rural-amenity landscapes, adjustments may be needed to allow for organising and sharing practices around ecologies that are heavily modified and novel in their composition. As we have shown, their modified and fragmented character presents different challenges from less modified landscapes.

While it is important to examine existing models for conservation collaboration, one notable contribution of this book is to show that regardless of the model, there must be a focus on sharing *practices*. It is sharing practices (rather than abstract knowledge) that can offer a rallying point for collaboration, around which people can pursue conservation goals that are attuned to landscape legacy and socio-ecological dynamism. The benefits of meetings and field days where practices can be shared intimately are well established (Meadows, Emtage, & Herbohn, 2014) and are particularly necessary in the case of conservation in rural-amenity landscapes, especially when we are arguing that practices are co-produced through human and nonhuman agency. This tangible understanding needs to be felt and experienced, and field days allow landholders to share their experiential learning in situ (Jones, 2019). The power of these affective experiences is clearly illustrated in Chapter 3.

The other benefit of sharing practices is the ability to develop a collective repository of conservation practice that can help monitor outcomes. Documenting the types, numbers and locations of species present in rural-amenity landscapes would form an important part of this picture. However, in keeping with our practice focus, we suggest a need to also document the experience, affects, perceptions and illustrations of human-environment interactions. As has recently been highlighted, perceptions and experiences of conservation action should be fundamental components of even the most scientifically derived monitoring framework (Bennett, 2016). Using public, online platforms for sharing audio-visual records of the travails of practice over time may prove very relevant material for rural-amenity landholders, especially given the applied ways they are likely to be learning. Sharing through these platforms might also help collaborative networks of landholders to identify what indicators of ecological flourishing or decline are worth monitoring in their location, further building knowledge of ecologies at scales beyond the individual property.

The sharing of stories of practice through a variety of mediums (much like those told by landholders here) will also be important for cultivating a notion of conservation practice that is changeable, responsive and deeply tied to place. Being able to tell the stories of conservation practice over time through multiple storytellers may offer a way to expand the custodianship of land. This might also enable the transfer of custodianship of land when properties change hands, rather than only transferring the legal obligations that come with the property title. As with the suggestions above, the idea of being able to host legacy stories in the public domain offers greater potential for shared learning that resists the bounding effect of private property on experiential learning and stewardship dispositions.

Dynamic socio-ecological change compels us to share conservation practices of the kind we articulate here, as their inherent messiness challenges notions that conservation outcomes are easily controllable or predictable (West, Beilin, & Wagenaar, 2019). The messiness of unfolding practice is an especially vital notion to share amidst the legacies of rural-amenity landscapes. Yet, as Chapter 3 revealed, it must be done in a way that empowers landholders to engage with ecologies while acknowledging that there may be no simple, repeatable 'right thing to do', and care is required to avoid the kind of uncertainty that discourages landholder efforts. Capturing, illustrating and storying practices in a readily accessible form would allow reflection on how practices are working (or not) at the scale of ecological processes and interactions, rather than at the scale of property parcels. It could also serve to galvanise solidarity around the collective challenge that is faced by everyone—something that many landholders in Chapter 6 would have appreciated. Collaboration at scales above the individual property might help to show, for example, that different conservation practices emerge in different places, given unique legacies and their influence on the way ecologies are being remade.

It is through sharing practices that more dynamic, reflexive and place-based goals for conservation in rural-amenity landscapes will emerge. With respect to plants, conservation goals might be sensitised to the capacities of particular plants (and their inter-relationships) and whether they serve to increase the diversity and habitat complexity of ecologies in a given space. As we have seen, there is a great deal of uncertainty in knowing which species will flourish and in what arrangement, and how the experiences of people who inhabit these landscapes will contribute to that flourishing. In practical terms, this might mean recognising that a single plant species might be the subject of preservation in one place and eradication in another, depending on species interrelationships.

Through practice-sharing, conservation goals should be orientated towards enrolling people collectively in the lives of plants, in pursuit of species diversity and abundance. While there is evidence that such nuanced conservation interventions play out on the ground (Head, Atchison, & Phillips, 2014), as yet this has not led to substantial shifts in discourse, objective and policy around the goals of conservation interventions. By adopting goals in rural-amenity landscapes that acknowledge landscape legacy, conservation practice becomes about encouraging people to work actively within the trajectories of ecological assemblages to build on existing species diversity. The capacity for landholders to be sensitised to nonhuman agency through the transgressive mobility of plants across the landscape suggests that plants could be an anchoring point for conservation goals attuned to dynamic change and collective responsibility.

2 Re-imagining Private Property Relations Amidst Colonial Conservation

Building from our introduction of nonhuman agency in Chapter 1, an embrace of nonhuman agency foregrounds Indigenous worldviews as part of any re-imagining of the governance structures around private land conservation. Indeed, there is substantial pressure being applied through international conservation organisations and agreements—primarily through the tireless action of First Nations Peoples—that is compelling nations like Australia to take the implications of settler-colonial conservation seriously (Stevens, 2014). The integration of key principles of the *Indigenous Peoples' Declaration to the World Parks Congress* (2003) into global conservation agreements and the adoption of the United Nations Declaration on the Rights of Indigenous Peoples (UNDRIP) by the International Union for Conservation of Nature (IUCN) are indicative of shifting international perspectives on protected areas. The fact that the World Commission on Protected Areas (WCPA) has targets and actions aimed at redressing land theft and past injustices inflicted on First Nations Peoples as part of its conservation objectives compels settler-colonial nations to respond meaningfully. What is critical to note here is that while public lands have been a focal point in exposing colonial conservation, international organisations like the IUCN have made clear that equity, justice and land rights issues for Indigenous Peoples are not just issues for public land, but for all land tenure types (IUCN, 2008).

In Australia, it is clear that the goals and settings for private land conservation need to challenge the past and present of settler-colonialism. In settler-colonial nations, questions of how to reconfigure private property relations to acknowledge nonhuman agency must attend to the rights, world view and land-use practices of First Nations. The conservation movement more broadly must challenge a historic and continuing entanglement with settler-colonialism, as efforts to reconfigure institutions and power structures for more equitable purposes in settler-colonial states will only re-escribe injustices if First Nations Peoples are excluded (Coulthard, 2014).

This is especially true in Australia when we are talking about conservation on lands that were never ceded to colonial invaders, and for which the nation-state has no treaty with First Nations.

For this reason, we argue that any re-imagining of property relations must also address settler-colonial relationships to land. As Ngurra et al. (2019) note, when it comes to land management and conservation, token acknowledgements of Indigenous knowledge are desperately inadequate. We need to consider what it means to "be led by Country… [remembering that] places also care, and care-full relationships honour and assert Country's agencies" (Ngurra et al., 2019, p. 12). The experiences of landholders in this book demonstrate how western, liberal property systems can facilitate certain forms of human-environment relations, but foreclose others. While we saw how landholders developed intimate conservation practices in collaboration with plants, these were often undergirded by individualistic notions of ownership. For conservation practices to flourish in diverse and contingent ways in the face of climate change—and in ways that acknowledge the terms of nonhuman life—we must consider how private property rights could be organised differently.

One way of shifting property relations in a way that encourages obligations to nonhumans and recognises unceded Indigenous lands could be to reconceive of property rights as held *jointly* by humans and nonhumans (Brown, Flemsæter, & Rønningen, 2019). A relevant example is the idea of a right of movement across the landscape for South Sámi herders and reindeers in Norway (Brown et al., 2019, p. 61). In this particular example, the current and future needs for reindeer and their South Sámi human herders to migrate across the landscape in response to a changing climate are framed as a relational unit that could be allocated a legal property right. This is an explicit attempt to imagine the spatial organisation of the landscape in a way that responds to the mobility of ecologies on scales that are relevant to their persistence, going beyond purely human terms. A property right that is allocated to a caring relationship between humans and particular plants or vegetation communities speaks directly to the need for collective conservation practices grounded in the spatial distribution of ecologies.

The idea of relational property rights like those outlined above is only of interest if they can contribute towards decolonising futures on the unceded lands of First Nations Peoples. In an Australian context, one area where a more-than-human property right could be examined is around caring-as-Country (Suchet-Pearson, Wright, Lloyd, & Burarrwanga, 2013) through cultural burning of ecologies by First Nations Peoples. First Nations Peoples in Australia are leading cultural burning practices in collaboration with government agencies that are challenging settler-colonial institutions' land management approaches (Neale, Carter, Nelson, & Bourke, 2019). Neale et al. (2019) discuss how cultural burning on Dja Dja Wurrung Country in Central Victoria is serving as a kind of 'decolonising experiment', progressing modest but tangible gains in the exercising of Indigenous authority over the way Country is cared for. Such experiments are pivotal in enabling further questions about the legal structures of the settler-colonial state that could not previously be posed. One such question is whether existing private property relations are constraining the conduct of caring-as-Country, and what avenues might be available for overcoming any constraints. As noted in The Victorian Traditional Owner Cultural Fire Strategy (2019), differing land tenures and property infrastructure like fences can present complications when it comes to cultural burning. The relatively small property sizes in many rural-amenity landscapes only exacerbates this issue. It is in this context of exploring how the settler-colonial state needs to respond to First Nations Lore and cultural responsibilities for burning Country that more-than-human property rights might offer some resonance.

While the idea of shifting dominant property relations might seem particularly challenging, it is worth remembering that property is not static and unchangeable. Property rights have shifted regularly over time—the un-bundling of water from land rights is but one recent example. The granting of legal personhood to rivers, like the Whanganui River in New Zealand and the Ganges in India, is an even more recent and radical example of shifting rights associated with nature. As we have emphasised, there is a real need to challenge existing property rights regimes in ways that enable a kind of "reconfiguring [of] responsibility" (Country et al., 2018, p. 4) for human-environment relations. The conservation practices of landholders in this book show us that people are often already immersed in many and varied caring relationships with plants, which sensitise them to notions of responsibility for more-than-human worlds. What is needed are ways of spatialising and organising the landscape that help to orientate conservation towards more collective processes of caring.

In concluding this book, we wish to highlight three main contributions to private land conservation discourse. First, conservation practices are multispecies achievements, shaped substantially by experiential learning and landscape legacies, and often centred on tangible interactions with plants. An awareness of practices helps to break down notions of human control and certainty over conservation outcomes and to reveal the importance of place-based, dynamic and reflexive goals for conservation. Secondly, we have shown that in the rural-amenity landscape context, where private land conservation challenges are heightened by ecological fragmentation and private property relations, there is a need for more collective conservation action at spatial and temporal scales beyond private property parcels. Recognising and sharing conservation practices across the landscape offers a basis on which to develop collective conservation goals that can respond to diverse ecologies in ways that acknowledge uncertainty and change over time.

Our third and final contribution is to show that a dominant relationship underpinning conservation practice is that between the isolated property owner and their delineated property territory. We show how property-centric, individual conservation practice renders invisible the complex relationality between and amongst species that produce ecologies over varying spatial and temporal scales. Collective goals for conservation across private land are vital, but will continue to be constrained unless the role of property relations in dictating those goals is addressed. We argue the need for different forms of property relations (or rights) that are capable of being shared by humans and nonhumans in ways that respond to First Nations sovereignty. This is a call for more inclusive forms of conservation that can meet our unfolding socio-ecological crisis in ways that enable and encourage collective action at spatial scales that are meaningful to the lives of nonhuman participants such as plants.

References

Bennett, N. J. (2016). Using perceptions as evidence to improve conservation and environmental management. *Conservation Biology, 30*(3), 582–592. https://doi.org/10.1111/cobi.12681.

Brown, K. M., Flemsæter, F., & Rønningen, K. (2019, January). More-than-human geographies of property: Moving towards spatial justice with response-ability. *Geoforum 99*, 54–62. https://doi.org/10.1016/j.geoforum.2018.12.012.

Coulthard, G. (2014). *Red skin, white masks: Rejecting the colonial politics of recognition*. http://www.upress.umn.edu/book-division/books/dorsality.

Country, B., Suchet-Pearson, S., Wright, S., Lloyd, K., Tofa, M., Sweeney, J., & Burarrwanga, L. (2018). Goŋ Gurtha: Enacting response-abilities as situated co-becoming Bawaka country. *Environment and Planning D: Society and Space*. https://doi.org/10.1177/0263775818799749.

Edwards, R., & Fox, T. (2012). Conservation Management Networks: The Gippsland grassy plains story. In J. Fitzsimons, I. Pulsford, & G. Wescott (Eds.), *Linking Australia's landscapes: Lessons and opportunities for large-scale conservation networks.* Melbourne: CSIRO.
Head, L., Atchison, J., & Phillips, C. (2014). The distinctive capacities of plants: Re-thinking difference via invasive species. *Transactions of the Institute of British Geographers, 40*(3), 399–413. https://doi.org/10.1111/tran.12077.
Hill, D. T., Collins, M. B., & Vidon, E. S. (2018, April). The environment and environmental justice: Linking the biophysical and the social using watershed boundaries. *Applied Geography 95*, 54–60. https://doi.org/10.1016/j.apgeog.2018.04.007.
IUCN. (2008, October). Selected resolutions and recommendations Indigenous peoples at the IUCN World Conservation Congress.
Jones, R. (2019). Really shit work? Bodily becoming and the capacity to care for the urban forest. *Social and Cultural Geography 20*(5), 630–648. https://doi.org/10.1080/14649365.2017.1384046.
Meadows, J., Emtage, N., & Herbohn, J. (2014). Engaging Australian small-scale lifestyle landowners in natural resource management programmes—Perceptions, past experiences and policy implications. *Land Use Policy* 36, 618–627. https://doi.org/10.1016/j.landusepol.2013.10.016.
Neale, T., Carter, R., Nelson, T., & Bourke, M. (2019). Walking together: A decolonising experiment in bushfire management on Dja Dja Wurrung Country. *Cultural Geographies.* https://doi.org/10.1177/1474474018821419.
Ngurra, D., Dadd, L., Glass, P., Scott, R., Graham, M., Judge, S., … Suchet-Pearson, S. (2019). Yanama Budyari Gumada: Reframing the urban to care as Darug Country in western Sydney. *Australian Geographer*, 1–15. https://doi.org/10.1080/00049182.2019.1601150.
Pratt, K. (2012). Rethinking community: Conservation, practice, and emotion. *Emotion, Space and Society, 5*(3), 177–185. https://doi.org/10.1016/j.emospa.2011.08.003.
Simpson, L. B. (2019). *As we have always done: Indigenous freedom through radical resistance.* Minneapolis: Minesota University Press.
Stevens, S. (2014). Indigenous peoples, biocultural diversity, and protected areas. In S. Stevens (Ed.), *Indigenous peoples, national parks, and protected areas: A new paradigm linking conservation, culture, and rights.* Tucson: University of Arizona Press.
Suchet-Pearson, S., Wright, S., Lloyd, K., & Burarrwanga, L. (2013). Caring as country: Towards an ontology of co-becoming in natural resource management. *Asia Pacific Viewpoint, 54*(2), 185–197. https://doi.org/10.1111/apv.12018.
Tennent, R, & Lockie, S. (2012, September). Vale landcare: The rise and decline of community-based natural resource management in rural Australia. *Journal of*

Environmental Planning and Management, 1–16. https://doi.org/10.1080/09640568.2012.689617.

The Victorian Traditional Owner Cultural Fire Strategy. (2019). *The victorian traditional owner cultural fire knowledge group*. Victoria, Australia. https://static1.squarespace.com/static/5b337bd52714e5a3a3f671e2/t/5cdb894beef1a16c0ac2b74d/1557891412693/Fire+strategy+final.pdf.

West, S, Beilin, R., & Wagenaar, H. (2019, July). Introducing a practice perspective on monitoring for adaptive management. *People and Nature*. https://doi.org/10.1002/pan3.10033.

Index

B. Cooke and R. Lane, *Making Ecologies on Private Land*,
https://doi.org/10.1007/978-3-030-31218-3

The manufacturer's authorised representative in the EU is Springer Nature Customer Service Centre GmbH, Europaplatz 3, 69115 Heidelberg, Germany. If you have any concerns regarding our products, please contact ProductSafety@springernature.com

Printed and bound by CPI Group (UK) Ltd, Croydon, CR0 4YY
05/10/2026
02228153-0001